EATCS
Monographs on Theoretical Computer Science
Volume 4

Wolfgang Reisig

PETRI NETS

An Introduction

With 111 Figures

1985

Springer-Verlag
Berlin Heidelberg New York Tokyo

Dr. Wolfgang Reisig
GMD
Postfach 12 40, Schloß Birlinghoven
5205 St. Augustin 1, Germany

Prof. Dr. Wilfried Brauer
FB Informatik der Universität
Rothenbaum-Chaussee 67–69, 2000 Hamburg 13, Germany

Prof. Dr. Grzegorz Rozenberg
Institut of Applied Mathematics and Computer Science
University of Leiden, Wassenaarseweg 80, P.O. Box 9512
2300 RA Leiden, The Netherlands

Prof. Dr. Arto Salomaa
Department of Mathematics, University of Turku
20500 Turku 50, Finland

Translation of the German original edition: W. Reisig, Petrinetze
ISBN 3-540-11478-5
Springer-Verlag Berlin Heidelberg New York 1982

ISBN 3-540-13723-8 Springer-Verlag Berlin Heidelberg New York Tokyo
ISBN 0-387-13723-8 Springer-Verlag New York Heidelberg Berlin Tokyo

Library of Congress Cataloging in Publication Data
Reisig, Wolfgang, 1950–
Petri nets.
Based on lectures given by the author at the Technical University of Aachen.
Translation of Petrinetze.
Includes index.
1. Petri nets. I. Title.
QA267.R4513 1985 511 84-26700
ISBN 0-387-13723-8 (U.S.)

Typesetting, printing and bookbinding: K. Triltsch, Würzburg
2145/3140-5 4 3 2 1 0

Preface

Net theory is a theory of systems organization which had its origins, about 20 years ago, in the dissertation of C. A. Petri [1]. Since this seminal paper, nets have been applied in various areas, at the same time being modified and theoretically investigated. In recent time, computer scientists are taking a broader interest in net theory.

The main concern of this book is the presentation of those parts of net theory which can serve as a basis for practical application. It introduces the basic net theoretical concepts and ways of thinking, motivates them by means of examples and derives relations between them. Some extended examples illustrate the method of application of nets. A major emphasis is devoted to those aspect which distinguish nets from other system models. These are for instance, the role of concurrency, an awareness of the finiteness of resources, and the possibility of using the same representation technique of different levels of abstraction. On completing this book the reader should have achieved a systematic grounding in the subject allowing him access to the net literature [25].

These objectives determined the subjects treated here.

The presentation of the material here is rather more axiomatic than inductive. We start with the basic notions of 'condition' and 'event' and the concept of the change of states by (concurrently) occurring events. By generalization of these notions a part of the theory of nets is presented. It would have been possible to proceed in the opposite order by firstly presenting net representations of practical, real systems and then, proceeding by a sequence of abstraction steps, reaching nets consisting of conditions and events. However, the chosen method of presentation corresponds to the usual way of proceeding in the framework of theoretical computer science.

It is not intended, in this book, to give a total overview and summary of the theory and applications of nets. Such an attempt is doomed to failure, not only because of the number of publications in the field, more than 500 are referenced in [25], but also because of the wide spectrum of the topics covered; for example complexity theory, the theory of formal languages, the theory and design of logic circuits, computer architecture, operating systems, the connection of computer processors, process control and real time systems, programming and command languages, databases, communication protocols, software engineering and yet even further into topics outside computer science (administration, jurisprudence, the logic of inter-personal interaction). Also, we are not able here to treat the foundations of net theory which lie in the philosophies of natural sciences, in the classical and non-classical logics, in theoretical physics and in the theories of communication.

A series of lectures for students in the third and fourth year of computer science, which the author gave at the Technical University of Aachen, served as a basis for this book. It might therefore be used in university courses, but it is also intended for the graduate student, the researcher and the professional who want to start within the field of Petri Nets.

The book assumes only an elementary knowledge of the structure, functioning and application of computer based information systems and some elementary mathematics. Using the first chapter as a basis, Part 1 and Part 2 may be read rather independently of each other. Part 3 uses the notions introduced in Chapts. 2, 4, 5 and 6. The computing practitioner should, in addition to the first chapter, find it worthwhile to study, particularly, the example at the start of Chapt. 5 and Sects. 6.3 to 6.5 and 8.1 to 8.3.

At the end of each chapter exercises are given. The more difficult ones are marked with *.

The appendix presents the mathematical notions and notation which are used in this book.

This book was originally published in German by Springer-Verlag in 1982. For the English edition it was revised and the "Further Reading" appendix and the exercises were incorporated.

Acknowledgements

This book could not have been created without the help of a number of people. At the Institut für Informationssystemforschung of the Gesellschaft für Mathematik und Datenverarbeitung in Bonn (West Germany), I received great support in discussing particular topics from C. A. Petri, H. Genrich, K. Lautenbach und P. S. Thiagarajan. Prof. W. Brauer gave many valuable remarks on the German manuscript.

On the occasion of the English translation it was possible to revise the text due to many hints and comments from its readers. Especially I am indepted to Eike Best, Ursula Goltz, Kurt Lautenbach, Roberto Minio, Horst Müller, Leo Ojala, Anastasia Pagnoni, Grzegorz Rozenberg and P.S. Thiagarajan for their many critical and constructive notes. Horst Müller and Dirk Hauschildt mainly contributed to the revision of Lemma 5.3 (d) and Theorem 7.2 (k), respectively.

I am deeply indebted to the translators Ursula Goltz and Dan Simpson, who with remarkable competence, fervour and patience did an excellent job. They also brought up some valuable discussion with regard to the contents of the book.

W.R.
Aachen, Germany
June 1983

To the English Edition

We have retained the notation of the German book (e.g. B for sets of conditions and S for sets of places) corresponding to the standards introduced at the Advanced Course on Net Theory and Application, cf. [17]. Any changes might have induced further problems (e.g. C for conditions would exclude an appropriate notation for cases. P for places would imply the non-standard notion of P-invariant).

U.G., D.S., Aachen and Sheffield

Contents

Introduction

(a) *Petri nets*, the subject of this book, are a model for procedures, organizations and devices where regulated flows, in particular information flows, play a role.

This language of nets arose from the intention of devising a *conceptual* and theoretical basis "for the description, in a uniform and exact manner, of as great as possible a number of phenomena related to *information transmission and information transformation*" [1]. We shall restrict ourselves to such applications of this theory as lie in the area of the design and use of computer based information systems.

In comparison with other system models, the major characteristics of Petri nets are the following:

- Causal dependencies and independencies in some set of events may be represented explicitly. Events which are independent of each other are not projected onto a linear timescale; instead, a non-interleaving, partial order relation of *concurrency* is introduced. This relation is fundamental for the whole conceptual basis of net theory.
- For some systems it may not be sensible to try to describe them as sequential functions. To do so only leads into unnecessary distracting detail. Examples are a query answering system of a distributed database, a real time system for production control, the control of processes in an operating system or a communication protocol.
- Systems may be represented at different *levels of abstraction* without having to change the description language. These levels of abstraction range from the change of single bits in computer memories to the embedding of a computer system into its environment.
- Net representations make it possible to *verify system properties* and to do correctness proofs in a specific way. Once a system has been modelled as a net, properties of the system may be represented by similar means, and correctness proofs may be built using the methods of net theory. Logical propositions are obtained as static components of dynamic net models.

Two objections may be raised here. One is that other methods which are well-known and established aim for the same goals. The other point is made by considering processes which run independently of each other (for example: processes in the central memory and in peripheral processing units of some computer). Such processes take particular states and perform state changes. The argument is that such states or changes which are coincidential may be combined into a global state or a global state change which covers these. Thus, a new theory is not required. Here we are not able to discuss in full the reasons why the specific ways of thinking of net theory are sufficiently important to justify the construction of a whole new theory. We simply note two points in reply: first, that the above proposed combination of coincident states or changes gives rise to the problem of determining whether they are really simultaneous. Secondly, a purely sequential model does not truly reflect the real *causal structure* of processes. In any sequentializing view we can not differentiate whether two events occur one after the other because the first is a prerequisite of the second or whether this order in time is solely by chance. But, in fact, the causal relations are those which, to a large extent, characterize a system.

(b) In the first chapter we shall present, by means of several examples, different net models. This gives a first insight into the structural patterns and representation methods typical for nets. The mathematically oriented reader may start at 1.1 and skip to 1.5.

Systems consisting of *conditions and events,* which are introduced in Part 1 of this book, constitute the most detailed description level of marked nets. Here, the fundamental notions of non-sequential processes are studied: viz., the relations of causal dependency and independency of events; the relationship between non-sequential processes and their set of possible sequential realizations; the metric of synchronic distances as a measure for the dependency between events; and, finally, the formulation of system properties in the language of logic and their integration into the net calculus.

In the second part of the book we consider nets consisting of *places and transitions.* Such nets are particularly suited to the formulation of blocking problems. For the investigation of such nets we introduce coverability graphs, which allow conclusions to be drawn about the behaviour of the system we are modelling.

We concentrate our presentation on those investigation methods which do not rely on the set of all possible sequential executions. A particular one of these is the calculus of invariants involving linear algebraic techniques. By means of several examples we show how this calculus may be used for the verification of system properties. For particular place/transition-nets, we derive particular methods of analysis.

In the third part of the book we consider *individuals, predicates* and *relations* on nets; we thus reach a level which yields a relationship between nets and universal algebra. We show how, on this level too, system properties which are formulated in the language of logic may yet again be represented in the net calculus. The verification of system properties so represented is again aided by an invariant calculus generalized from place/transition-nets.

Chapter 1

Introductory Examples and Basic Definitions

1.1 Examples from Different Areas

In the preface and the introduction, we have already used the terms "system organization", "system model", "condition", "event" and "information transformation" without explaining them. These notions are of fundamental importance in net theory. However, as they are concepts from the real world, we shall not try to give precise definitions of them but rather appeal to the intuition and general understanding of the reader. But, we shall have to consider properties of objects of this kind, and also the relationships between such objects. We shall say, for instance, that "system models" represent real systems more or less adequately, that "events" occur and that "conditions" do or do not hold.

(a) Let us first consider systems comprising *conditions* and *events*. Figure 1 shows a system in which the conditions are: "it is spring", "it is summer", "it is autumn" and "it is winter"; the events are: "start of spring", "start of summer", "start of autumn" and "start of winter". We see that each condition is represented by a circle and each event by a box. Each condition which holds is marked by a dot (a *token*) (in Fig. 1, it is "spring"). The set of conditions which hold in some configuration is called a *case*. In the system represented in

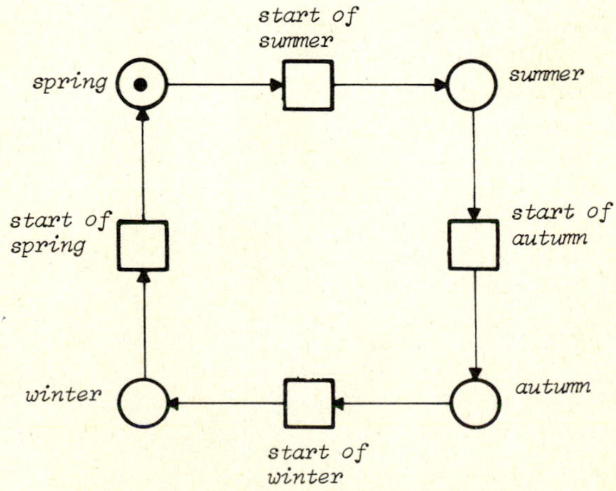

Fig. 1. The four seasons and their changes

Fig. 1, each case has only one element. Whenever an event occurs, another case results. A condition, *b*, and an event, *e*, may be related with each other as follows:

(1) *b* starts to hold when *e* occurs. *b* is then called a *postcondition* of *e*. Graphically, this relationship is represented as an arc from *e* to *b*.
(2) *b* ceases to hold when *e* occurs. *b* is then called a *precondition* of *e*. Graphically, this relationship is represented as an arc from *b* to *e*.

If *b* is not affected by the occurrence of *e* there is no arc between *b* and *e* at all.

So, in our system of the four seasons, when an event occurs the token is moved to the next season.

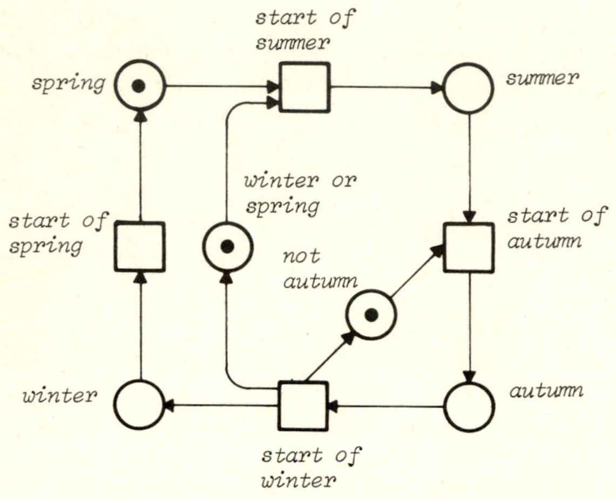

Fig. 2. Addition of two conditions to Fig. 1

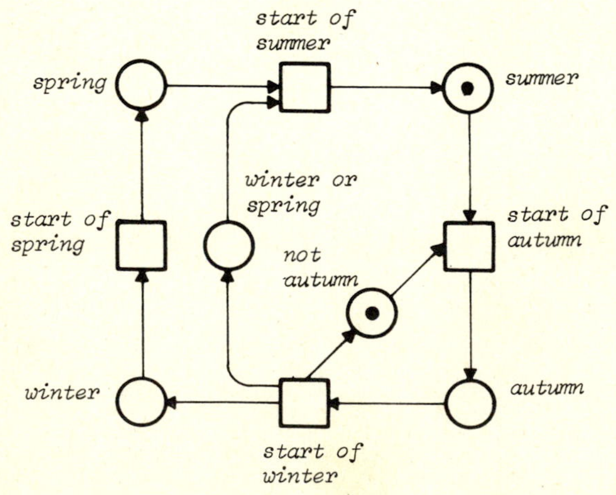

Fig. 3. The system of Fig. 2 after start of summer

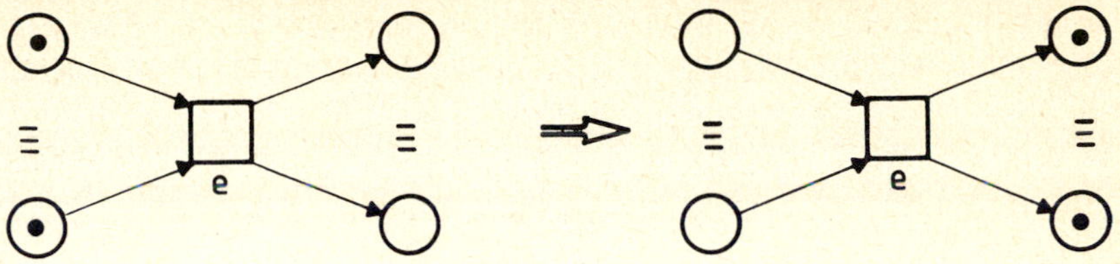

Fig. 4. The occurrence of an event *e*

When modelling the four seasons and their changes we may wish to represent additional conditions and events. When we add the conditions "winter or spring" and "not autumn", we obtain the system shown in Fig. 2. Note that now some events have several pre- or postconditions.

In the system represented in Fig. 2, consider now in which case the event "start of summer" may occur. This is when it is both "spring" and "winter or spring", and it is not already "summer". By the occurrence of this event we obtain the configuration shown in Fig. 3. In general, an event may occur if all its preconditions hold and none of its postconditions hold. Figure 4 shows the requirements for, and the result of, an event, *e*, occuring.

Although it is certainly an interesting event that winter ends, it should not be distinguished from the start of spring because neither of these events can occur without the other. The end of winter and the start of spring are *coincident* events, they are represented by one single box.

(b) When describing systems, at some levels, it is not always appropriate to use the notions of "condition" and "event". For example, when looking for

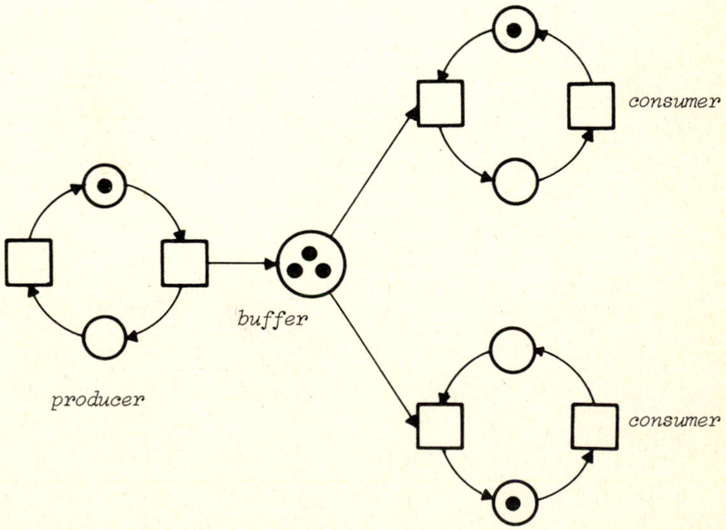

Fig. 5. A system consisting of one producer and two consumers

bottlenecks in manufacturing processes, it may only be the total number of goods produced which is of interest and not their individual identities. In the representation of a store, a set of conditions ("the places $s_1, \ldots, s_n$ are used") may then be combined into one item which is marked n-times ("n places are used"). Figure 5 shows a system of one producer and two consumers using a buffer as their store. The producer generates items (represented as tokens), which are placed in the buffer. The consumers may remove items which are in the buffer. In such nets, we say the elements are *places* ○ and *transitions* □. Places may, in contrast to conditions, carry more than one token. Arcs again indicate the flow of tokens. A transition *fires* by removing a token from each *input place* and by adding a token to each *output place* (Fig. 6). If we restrict each place to carry at most one token this firing rule corresponds to the effect of event occurrences described above.

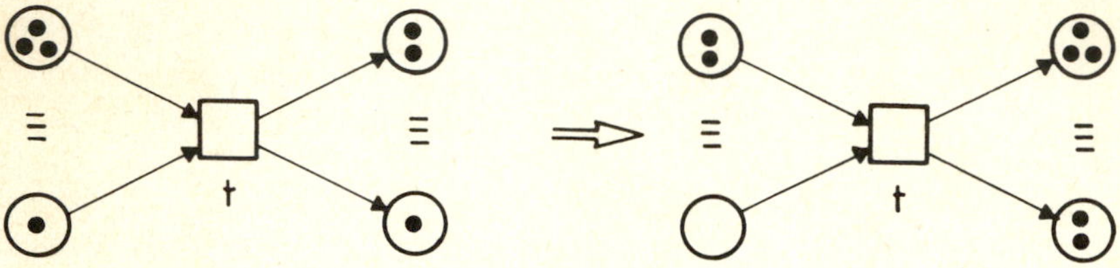

Fig. 6. The firing of a transition t

The two consumers are represented by two tokens in one single consumer part of the net as in Fig. 7. However, now the consumers may no longer be distinguished as individuals.

(c) Nets consisting of places and transitions model system properties concerning the number, the distribution and the flow of objects which are not further distinguished. If we wish to consider individual properties of the objects we must be able to identify particular tokens. Figure 8 shows a fragment of an industrial production system, the operation of which is intuitively clear. This also illustrates the construction of nets. Round nodes (places) represent passive system components. These are those components which may store items, take particular states and make things observable. Rectangular nodes (transitions)

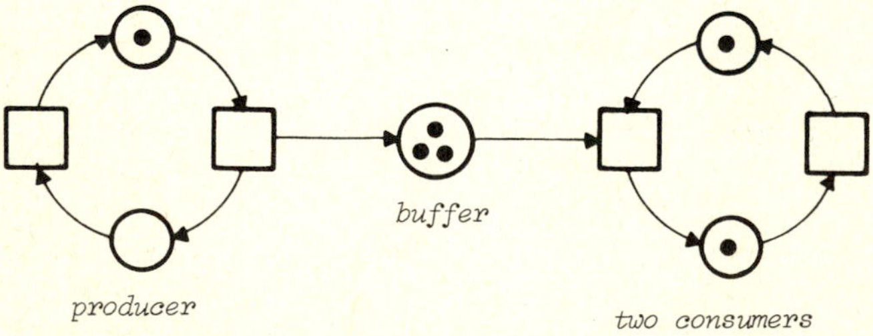

Fig. 7. Combination of the two consumers of Fig. 5 into one part of the net

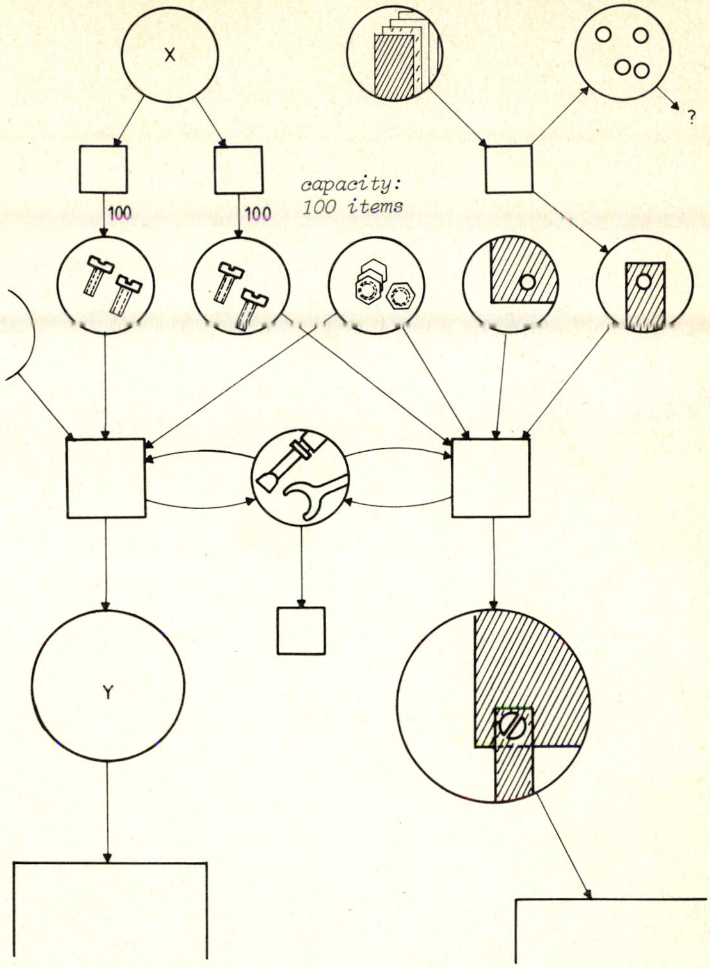

Fig. 8. A fragment of an industrial production system

represent active system components. Such components may produce, transport and change objects. Arcs show which system components are directly coupled with each other and in which direction objects may "flow" through the net. These objects themselves are represented as *individual tokens.*

(d) There are systems where some of the connections between system components are not oriented. Some systems do not have objects which flow. But we shall always adhere to the principle of partitioning the system into active and passive components. This partitioning may often be done in a number of different ways. For example, as a first approximation, a game of chess may be represented as an interaction, t, of two players (holders of states) s_1 and s_2. Alternatively, the board, s, may be considered as a passive object to be accessed by the moves, t_1 and t_2, made by each player. Figure 9 shows the first

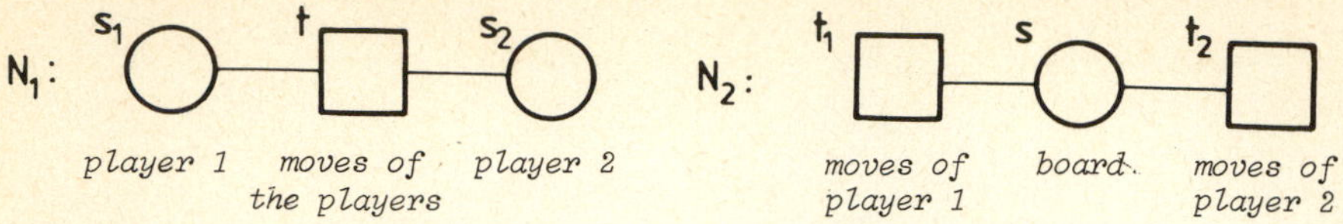

Fig. 9. Two representations of a chess game

view as N_1 and the second view as N_2. These two different partitions stress two different aspects of the same system. Each may be refined so that the aspects of the other view are included. Figure 10 shows the smallest refinement which covers the aspects of both views.

As long as no distinguished flow of objects is to be represented, the arcs of a net may be undirected, as in Fig. 9 and Fig. 10. In this book we will not discuss nets of this kind.

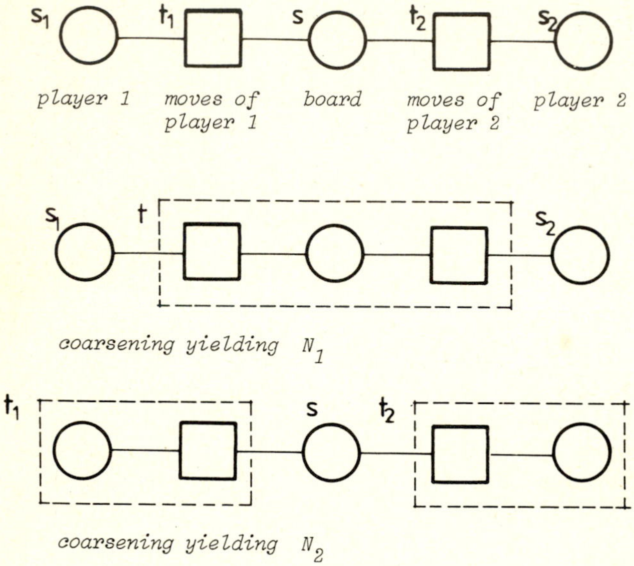

Fig. 10. Common refinement from Fig. 9

1.2 Examples from Logic Circuits and Operating Systems

(a) Let us start with a problem from logic circuits. x and y are two variables, which can take the values "true" and "false". Each is assigned an initial value independently of the other. They are then combined to give the value $x \wedge y$ to the variable x and the value $x \vee y$ to the variable y. These new values are available until they are, again independently, deleted. Then the system returns to the initial configuration and the variables may be given new values. Figure 11 shows this system as a net consisting of conditions and events.

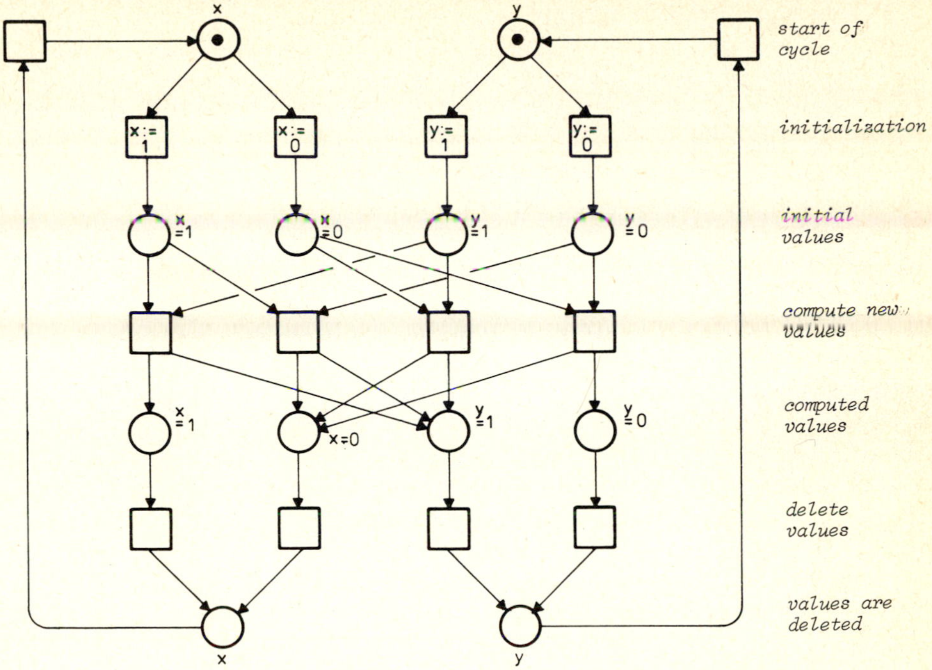

Fig. 11. A system which calculates $x := x \wedge y$ *and* $y := x \vee y$

(b) In operating systems several processes may write to, or read from, an area in main memory. For example, consider a configuration of two processes with write access and four processes with read access. At most three reader processes may overlap in their access to the memory. When the memory is being changed by some writer process no other process may have access.

Figure 12 shows this system as a net consisting of places and transitions. Two of the arcs are labelled by 3. In this case, when the appropriate transition fires the token count on the place s is reduced or increased by 3 instead of 1.

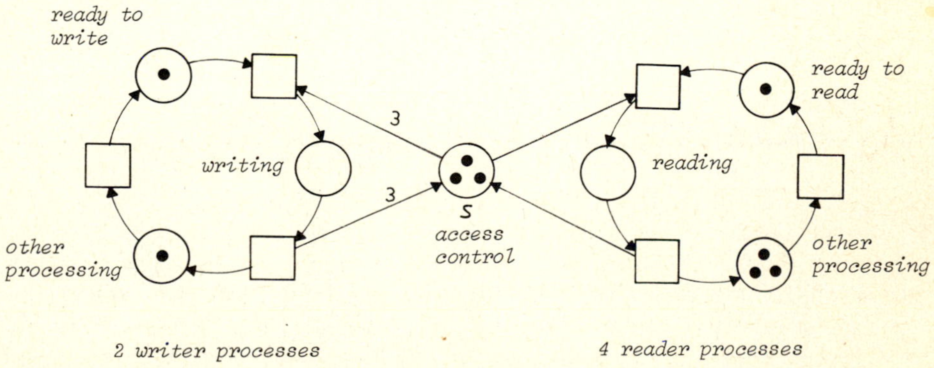

Fig. 12. Organization of the access rights of six processes to a memory area

1.3 Non-Sequential Programs

In the area of software engineering, non-sequential programs are required, in particular for systems programming and process control. Even in small programming problems the actual requirements for the program can not always be represented using purely sequential techniques. To do so means we must accept an overspecification. Because of the currently available computer architectures this overspecification may be economically advisable and seem more efficient, for if we specify the non-sequential behaviour we still have to give a sequential specification for the implementation. However, we propose it is a fundamental advantage to avoid the introduction of orderings except in those situations where they are necessary or wanted. To show this, we now consider two examples.

(a) We want to construct a program for the addition of two natural numbers stored in the variables x and y. In the final state, the variable x should contain the value 0 and the variable y the required sum. The operations allowed are the addition and the subtraction of the value 1 and the test for 0.

Figure 13 presents two sequential solutions to this problem. The nets shown there are similar to ordinary flow charts. Instructions are represented as events, and possible program states as conditions. The current state is marked by a token. In both nets, each event has exactly one pre- and exactly one postcondition. Hence, from the firing rule given above, there is always only one token in the net.

The two programs shown in Fig. 13 are almost identical. They differ only in the order of the instructions $x := x - 1$ and $y := y + 1$. Clearly their order is of no importance; actually, when executing them no order need be observed at all as they are logically independent.

Figure 14 shows a non-sequential program for the addition problem. Here, e_1 and e_2 change the number of tokens in the net from one to two and back to

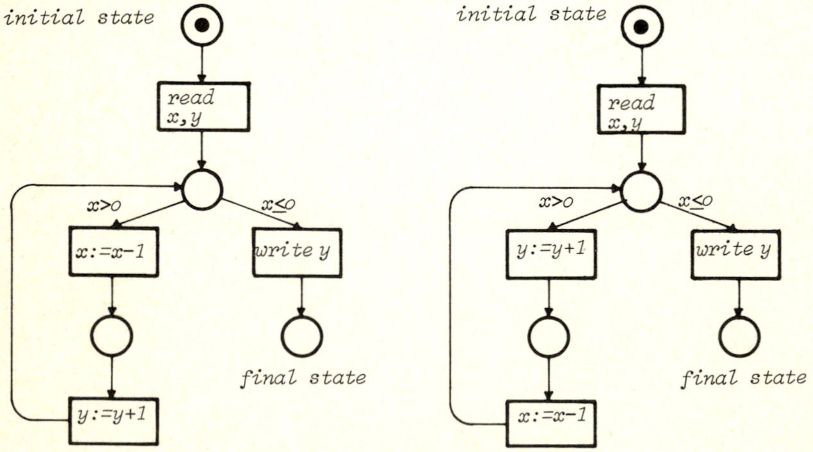

Fig. 13. Two sequential programs for the addition problem

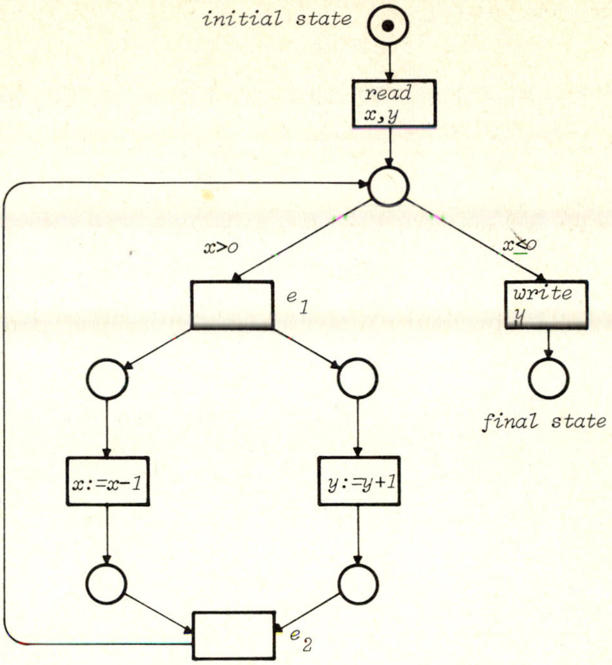

Fig. 14. A non-sequential program for the addition problem

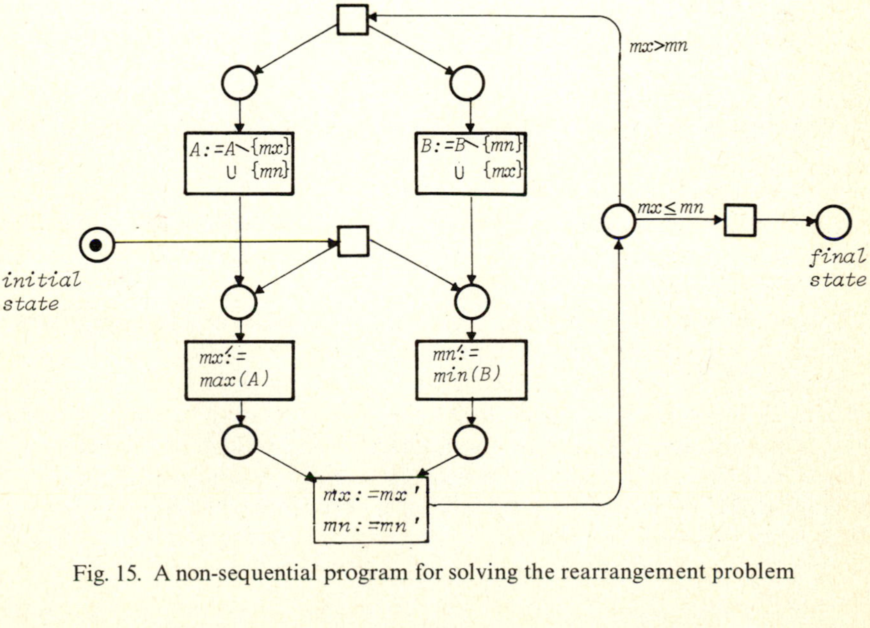

Fig. 15. A non-sequential program for solving the rearrangement problem

one, respectively. In this representation it is explicit that the instructions $x := x - 1$ and $y := y + 1$ may be executed independently.

(b) The program shown in Fig. 15 solves the following rearrangement problem: given two finite, non-empty, disjoint sets A and B, $A \cup B$ is to be rearranged into two subsets A' and B' such that $|A'| = |A|$, $|B'| = |B|$ and $\max(A') < \min(B')$.

Operations on sets are certainly slow in comparison with simple assignments. In this solution the set operations are executed concurrently whenever possible.

Non-sequential programs of the kind discussed in this section are not to be considered as special non-deterministic programs. In some particular run of such programs, it need not be decided in which order concurrent instructions are executed. The program is deterministic in that its meaning is independent of any ordering of instruction evaluations. If, nevertheless, a computing system chooses to impose an order, it performs a service which is beyond the requirements specified in the program.

1.4 An Example for Systems Analysis

Whenever computers are used for practical applications we have to develop programs from informal problem descriptions. Nets may be used to support this development in the following way. To start with, some structural properties are imposed on the informal description by some net representation. Then a series of gradual refinement steps follows, finally yielding system parts in a form suitable for programming. By this continuous and systematic development, we also obtain a description of how the system parts relate to each other and to their environment.

For example, consider the organization of the borrowing and the returning of books in a library. An unambigous and perspicuous representation of this organization is needed to describe several different levels and several different views of the organization. Different views correspond, for instance, to the needs of the library staff, the users, the suppliers of new books, the caretaker, the administration, etc.

Moreover, when the library system is set up, the designers of the library would need a representation of their view as would the designers of a computer-aided administration system.

Figure 16 shows a first coarse structuring of the library system. Users can access the library by three desks; the request desk, the collection desk and the return desk. In the library all books are kept in the stack and each book has an index card. A potential borrower enters the library system at the request desk where a particular book may be requested. If the book is in the library it is taken from the stack and the borrowed book index is updated. The user gets the book at the collection desk. When a user returns a book he does so via the return desk; the book is put back in the stack and the index is appropriately updated.

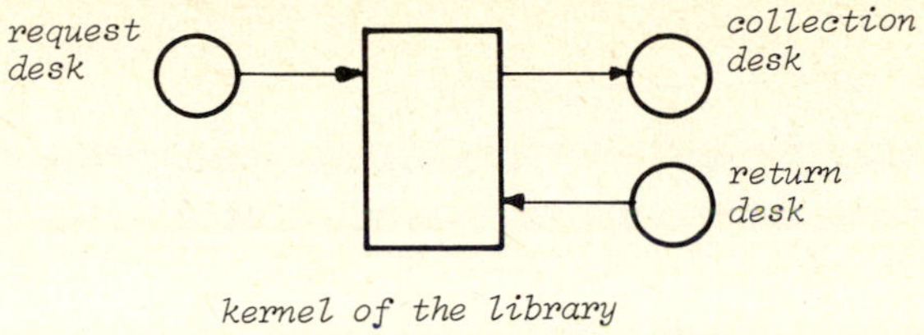

Fig. 16. Coarse structure of a library

A first refinement of Fig. 16, shown in Fig. 17 involves two active agencies which organize the delivery and re-acceptance of the books and two passive components, the stack and the borrowed book index.

Figure 18 represents a simple organization scheme for such a library. The tokens in this net are of three kinds: order forms, books and index cards. Each book is identified by a number and, for each book, there is an index card bearing this number. To borrow a book an order form containing the book number is put on the request desk. The book and its index card are taken from the stack, the book and the order form are placed on the collection desk and the index card is inserted in the borrowed book index. However, if the book requested has already been borrowed, the order form with an appropriate message is given to the collection desk. When a book is returned, the book together with its index card is replaced into the stack. Figure 18 illustrates a typical situation in this small library. Book *1* has been ordered and the corresponding order form is on the request desk. On the collection desk is book *3* with its order form, and also an order form saying book *5* is already borrowed. Book *2* has been returned and is still on the desk. The stack contains books *1* and *4* with their index cards and books *2*, *3* and *5* are borrowed.

A change to a new situation is possible by the occurrence of one of the three events e_1, e_2, e_3. For such an occurrence the objects written on the arcs

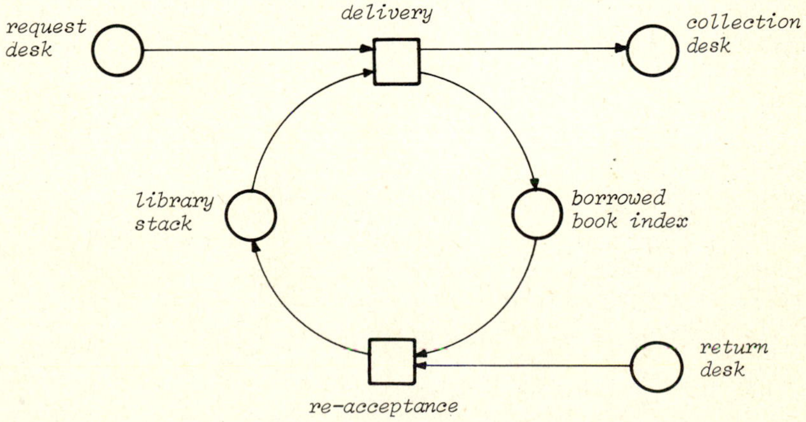

Fig. 17. Refinement of Fig. 16

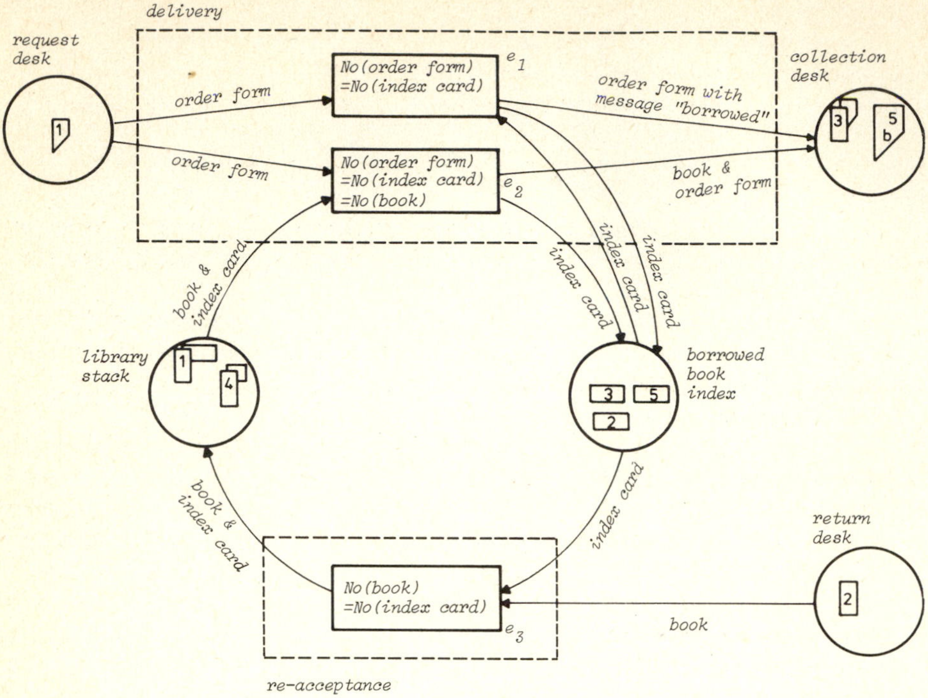

Fig. 18. A simple library organization

leading to e_i have to be instantiated by suitable items from the input places; the items must be chosen so that the formula written on e_i besomes true. When e_i occurs the corresponding tokens flow in accordance with the event occurrence rule for nets consisting of conditions and events. Such nets are called *predicate/events-nets.*

1.5 Some Basic Definitions

In all the constructions described in the previous sections, the underlying structure consisted of two sorts of objects and some relations between them. As long as these objects are not interpreted in any special way (for instance, as conditions, states, stores or events, state changes, instructions), we call circles *S-elements* and boxes *T-elements,* respectively*. The relations between *S*- and *T*-elements, represented as arcs, are combined yielding the *flow relation.* It was not accidental that the flow relation in the previous examples always connected elements of different sort. Rather, this is a basic property of nets.

* These terms are derived from the interpretation as places (German: Stellen) and transitions. To date, this has been the most thoroughly investigated and frequently applied interpretation.

We now make this precise in the following definition:

(a) Definition. A triple $N = (S, T; F)$ is called a *net* iff
(i) S and T are disjoint sets (the elements of S are called *S-elements,* the elements of T are called *T-elements*),
(ii) $F \subseteq (S \times T) \cup (T \times S)$ is a binary relation, the *flow relation* of N.

Graphically, we represent S-elements as circles and T-elements as boxes (mnemonically Ⓢ, 🅃). The flow relation is represented by arcs between the respective circles and boxes.

(b) Notation. Let $N = (S, T; F)$ be a net. We sometimes denote the three components S, T and F by S_N, T_N and F_N, respectively. If confusion can be excluded, we also write N for $S \cup T$.

(c) Definition. Let N be a net.
(i) For $x \in N$,
$^{\bullet}x = \{y \mid y F_N x\}$ is called the *preset* of x,
$x^{\bullet} = \{y \mid x F_N y\}$ is called the *postset* of x.
For $X \subseteq N$, let $^{\bullet}X = \bigcup_{x \in X} {}^{\bullet}x$ and $x^{\bullet} = \bigcup_{x \in X} x^{\bullet}$.

In particular we have, for $x, y \in N$:

$x \in {}^{\bullet}y \Leftrightarrow y \in x^{\bullet}$.

(ii) A pair $(s, t) \in S_N \times T_N$ is called a *self-loop* iff $s F_N t \wedge t F_N s$. N is called *pure* iff F_N does not contain any self-loops.
(iii) $x \in N$ is called *isolated* iff $^{\bullet}x \cup x^{\bullet} = \emptyset$.
(iv) N is called *simple* iff distinct elements do not have the same pre- and postset, i.e.

$\forall x, y \in N$: $(^{\bullet}x = {}^{\bullet}y \wedge x^{\bullet} = y^{\bullet}) \Rightarrow x = y$.

Figure 19 shows a net which is simple but not pure and which contains no isolated elements.

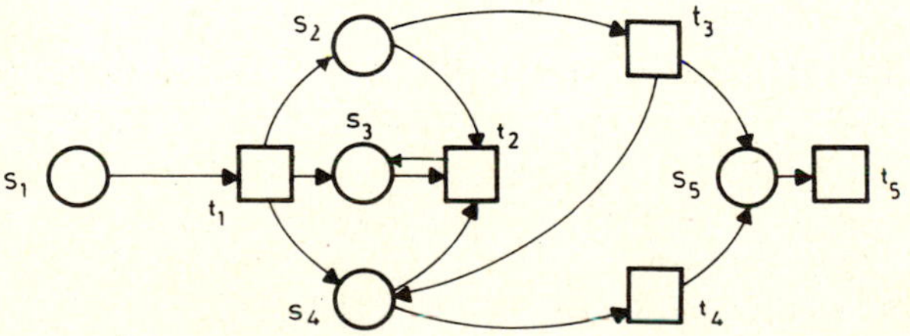

$S=\{s_1,\ldots,s_5\}$
$T=\{t_1,\ldots,t_5\}$
$F=\{(s_1,t_1),(t_1,s_2),(t_1,s_3),(t_1,s_4),(s_3,t_2),(t_2,s_3),(s_2,t_2),(s_2,t_3),(s_4,t_2),(s_4,t_4)$
$(t_3,s_4),(t_3,s_5),(t_4,s_5),(s_5,t_5)\}$

Fig. 19. Graphical representation of a net

(d) Definition. Let N and N' be two nets.

(i) Given a bijection β: $N \to N'$, we call N and N' *β-isomorphic* iff $s \in S_N \Leftrightarrow \beta(s) \in S_{N'}$ and $x F_N y \Leftrightarrow \beta(x) F_{N'} \beta(y)$. (This implies that $t \in T_N \Leftrightarrow \beta(t) \in T_{N'}$.)

(ii) N and N' are calle *isomorphic* iff they are β-isomorphic for some bijection β.

Graphical representations in which the elements are not named explicitly represent nets uniquely up to isomorphism. We always use such representations if the names of the elements are not important.

1.6 Summary and Overview

The examples given in this chapter may be classified into three groups: Nets consisting of conditions and events, nets consisting of places and transitions, and nets which carry individuals as tokens. Formally, these classes are distinguished mainly by the way the nets are marked. In the first case, an S-element is either marked or unmarked, in the second case it carries a certain number of indistinguishable tokens, in the third case it is marked by individual objects. The three parts of this book correspond to these three interpretations. Other interpretations (see, for instance, the chess game discussed in 1.1 (d)) will not be considered here.

Exercises for Chapter 1

1. Represent in Fig. 1 the two conditions
   a) "not winter and not spring",
   b) "spring or autumn".

2. Rearrange Fig. 12 so that in each case either none or more than one process is reading.

Part 1. Condition/Event-Systems

Part 1 deals with a fundamental class of systems in net theory called condition/event-systems. They are introduced in Chap. 2. In Chap. 3 we investigate what single processes running in such systems look like. Chapter 4 introduces and explains notions for the representation and description of some properties of condition/event-systems.

Chapter 2

Nets Consisting of Conditions and Events

First, for nets consisting of conditions and events, we must make precise the meaning of "occurrence of a single event or several independent events". For this, the notion of a *step* is introduced. A notion of equivalence for condition/event systems (C/E-systems) is then introduced, and we show how each system can be transformed to an equivalent *contact-free* normal form. Finally we discuss the *case graph* of a C/E-system. This graph provides an overview of all cases and steps of the system.

2.1 Cases and Steps

In the first chapter we have already informally discussed systems consisting of conditions and events. Conditions are represented as S-elements, events as T-elements. We know already that conditions are either satisfied or not, and that the occurrence of events changes condition holdings. In each configuration of such a system some conditions hold, while the rest do not hold. The set of conditions which hold in a configuration is called a *case*. An event e can occur in a case c, if and only if the preconditions of e belong to c and the postconditions of e do not belong to c. When e occurs, the preconditions of e cease to hold and the postconditions of e begin to hold.

If S- and T-elements are to be interpreted as conditions and events, respectively, we shall write $(B, E; F)$ instead of $(S, T; F)$.

(a) Definition. Let $N = (B, E; F)$ be a net.

(i) A subset $c \subseteq B$ is called a *case*.

(ii) Let $e \in E$ and $c \subseteq B$, e has *concession in c* (is *c-enabled*) iff ${}^{\bullet}e \subseteq c \wedge e^{\bullet} \cap c = \emptyset$.

(iii) Let $e \in E$, let $c \subseteq B$ and let e be c-enabled. $c' = (c \setminus {}^{\bullet}e) \cup e^{\bullet}$ is called the *follower case of c under e* (c' results from the *occurrence of e in the case c*) and we write: $c\,[e\rangle\, c'$.

To represent a case c graphically, we draw a dot (a *token*) in each circle belonging to c.

Figures 1 – 3 show nets consisting of conditions and events; one case is shown in each figure.

According to Definition (a), an event e can only occur if no condition in its postset $e^{\bullet}$ is satisfied. If any satisfied postconditions are preventing the occurrence of e; that is, if, in a case c, ${}^{\bullet}e \subseteq c \wedge e^{\bullet} \cap c \neq \emptyset$, then this is called a *con-tact-situation*. At first glance, it might not seem fully justified that e is then not allowed to occur: One could, for example, propose that every postcondition which is satisfied before the occurrence of e remains so after the occurrence of e. But let us discuss the implications. In terms of some examples, it would mean that spring may start when it is already spring; that an already written memory cell may be rewritten; that a full glass may be filled; that a reserved seat may be reserved again; or that a car may move to a place where another car is already standing. Some such events are impossible, but on the other hand some may be intended, or else possible but unwanted. We will see later how such events can be described, discovered or prevented. But at the lowest and most detailed level of description, which concerns us now, we rule them out. There are also formal reasons for this: Suppose that we allow a transition

⬤–▢–⬤ ⇒ ◯–▢–⬤ and that in the situation ⬤–▢–⬤–▢–◯

both events occur exactly once, then it depends on the order of their occurrences, whether the case ◯–▢–◯–▢–⬤ or the case ◯–▢–⬤–▢–⬤

results. But we want to be able to explicitly distinguish, to represent and to trace, whether events occur in a particular intended order or whether they occur in arbitrary order or independently.

When an event has led from one case to another, other events may occur, yielding yet other cases. These events are dependent on each other in different ways: In Fig. 20, for example, e_1 has to occur before e_3 and e_4. e_3 and e_4 on the one hand and e_2 on the other hand are alternatives. If e_3 and e_4 occur, they can be combined into one *step*. The occurrence of a set of events G in one step is possible if all events of G are enabled and their pre- and postsets are disjoint; G will then be called *detached*.

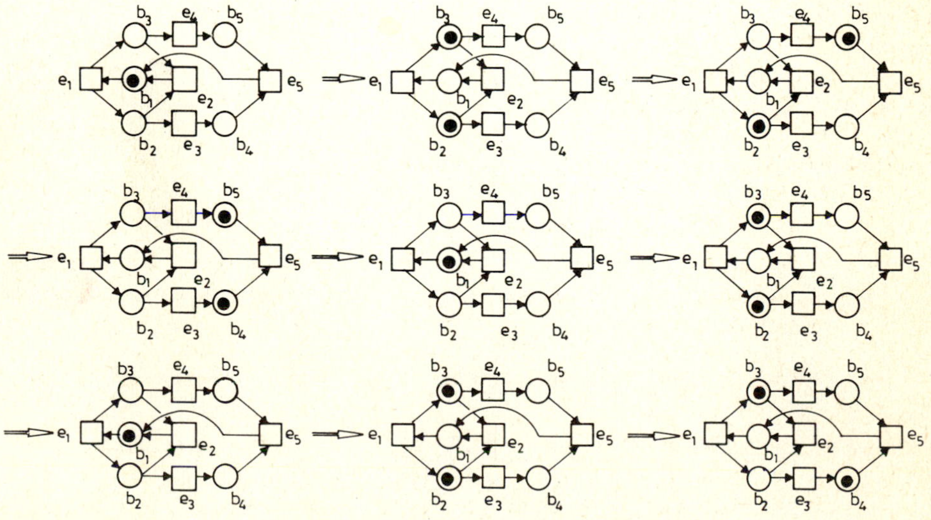

Fig. 20. The change of cases by event occurrences

(b) Definition. Let $N = (B, E; F)$ be a net.

(i) A set of events $G \subseteq E$ is called *detached* iff $\forall e_1, e_2 \in G: e_1 \neq e_2 \Rightarrow {}^\bullet e_1 \cap {}^\bullet e_2 = \emptyset = e_1^\bullet \cap e_2^\bullet$.

(ii) Let c and c' be cases of N and let G be detached. G is called a *step* from c to c' (notation: $c[G\rangle c'$) iff each event $e \in G$ is c-enabled and $c' = (c \setminus {}^\bullet G) \cup G^\bullet$.

By a step $c[G\rangle c'$, G leads from a case c to a case c'. Obviously, if G contains only one element, $G = \{e\}$: $c[G\rangle c' \Leftrightarrow c[e\rangle c'$.

The following lemma clarifies some relations between c, G and c'.

(c) Lemma. *Let N be a net, let $G \subseteq E_N$ be detached and let c_1, c_2* be cases of N. *Then*

$$c[G\rangle c' \Leftrightarrow c \setminus c' = {}^\bullet G \wedge c' \setminus c = G^\bullet.$$

Proof. If $c[G\rangle c'$, all $e \in G$ are enabled and $c' = (c \setminus {}^\bullet G) \cup G^\bullet$. Hence, ${}^\bullet G \subseteq c$ and $G^\bullet \cap c = \emptyset$.

Now it follows

$$\begin{aligned} c \setminus c' &= c \setminus ((c \setminus {}^\bullet G) \cup G^\bullet) \\ &= (c \setminus (c \setminus {}^\bullet G)) \cap (c \setminus G^\bullet) \text{ according to A3 (v) (cf. Appendix)} \\ &= (c \cap {}^\bullet G) \cap (c \setminus G^\bullet) \text{ according to A3 (ii)} \\ &= (c \cap {}^\bullet G) \text{ as } c \cap G^\bullet = \emptyset \\ &= {}^\bullet G \text{ as } {}^\bullet G \subseteq c. \end{aligned}$$

$$\begin{aligned} c' \setminus c &= ((c \setminus {}^\bullet G) \cup G^\bullet) \setminus c \\ &= ((c \setminus {}^\bullet G) \setminus c) \cup (G^\bullet \setminus c) \text{ according to A3 (iii)} \\ &= \emptyset \cup (G^\bullet \setminus c) \\ &= G^\bullet \text{ as } G^\bullet \cap c = \emptyset. \end{aligned}$$

Conversely, if $c \setminus c' = {}^\bullet G$ then ${}^\bullet G \subseteq c$, and if $c' \setminus c = G^\bullet$ then $G^\bullet \cap c = \emptyset$, hence all $e \in G$ are c-enabled. Furthermore,

$$\begin{aligned} (c \setminus {}^\bullet G) \cup G^\bullet &= (c \setminus (c \setminus c')) \cup (c' \setminus c) \\ &= (c \cap c') \cup (c' \setminus c) \text{ according to A3 (ii)} \\ &= c', \text{ hence } c[G\rangle c'. \end{aligned}$$

□

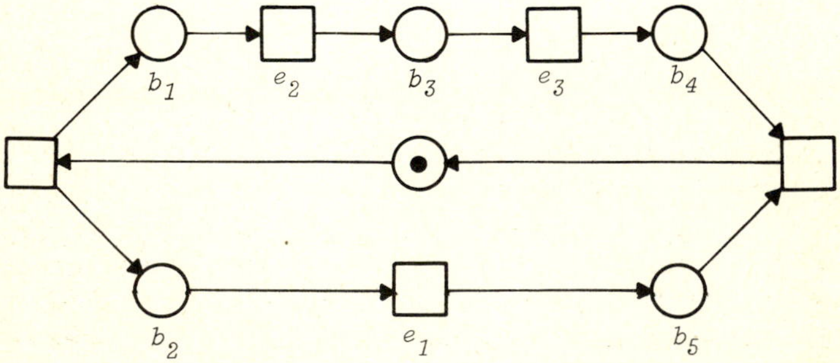

Fig. 21. $\{e_1, e_2\}$ is a step from $\{b_1, b_2\}$ to $\{b_3, b_5\}$, $\{e_1, e_3\}$ is a step from $\{b_2, b_3\}$ to $\{b_4, b_5\}$

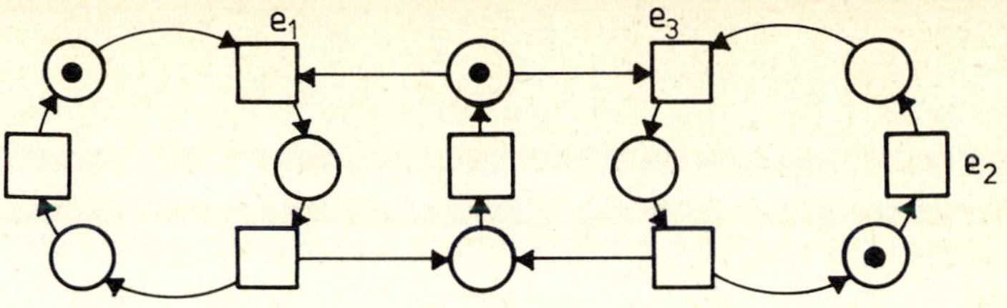

Fig. 22. A situation of confusion

In general there are several possibilities for combining events into steps: In Fig. 21 not only $\{e_1, e_2\}$ but also $\{e_1, e_3\}$ yield a step. By changing cases successively by steps, a *process* is generated (this notion will be made precise later).

If a step is finite, then it can be realized by the occurrence of its events in arbitrary order:

(d) Lemma. *Let N be a net, let c and c′ be cases of N and let G be a finite step from c to c′. Let* $(e_1, \ldots, e_n)$ *be an arbitrary ordering of the elements of G, such that* $G = \{e_1, \ldots, e_n\}$. *Then there are cases* $c_0, \ldots, c_n$, *such that* $c = c_0$, $c' = c_n$ *and* $c_{i-1}\,[e_i\rangle\, c_i$ $(i = 1, \ldots, n)$.

Proof. Let $e, e' \in G$ and let c be a case in which e and e' both have concession; then, ${}^\bullet e \cap {}^\bullet e' = \emptyset \wedge e^\bullet \cap {}^\bullet e' = \emptyset$. So, if $c\,[e\rangle\, c'$, ${}^\bullet e' \subseteq c$.

Analogously it can be shown that $e'^\bullet \cap c' = \emptyset$. So e' has concession in c'. For $i = 1, \ldots, n$ it follows that e_i remains activated during successive occurrence of $e_1, \ldots, e_{i-1}$ and can therefore transform c_{i-1} into c_i. □

It may be the case that two enabled events can occur in a single step. However it may be the case that they have common pre- or postconditions and that their occurrences are therefore mutually exclusive. Such events are said to be in *conflict* with each other. It may not be obvious whether conflicts will arise; for example, if in Fig. 22 starting with the case shown there, e_1 occurs before e_2 then there will be no conflict between e_1 and e_3. If, however, e_2 occurs before e_1 then such a conflict results. There is no order specified between e_1 and e_2; this is a situation called *confusion*.

2.2 Condition/Event-Systems

We will now introduce nets which model the notions of condition and event and which are intended to make derived notions, such as case and step, usable for the description of real systems.

A system consisting of conditions and events is not fully described until we specify, in addition to the net $(B, E; F)$, also the cases we wish to consider. (For example, the net of the four seasons in Fig. 1 would not make sense as intended with a case containing two elements.) Such a set of cases C should have the following properties:

1) If a step $G \subseteq E$ is possible in a case $c \in C$, then G leads again to a case in C (steps do not lead out of C).
2) Conversely, if a case $c \in C$ can result from a step $G \subseteq E$, then the situation we moved from was also a case of C. (In other words, when we reason backwards and look for preceding cases, we only find cases of C).
3) All cases in C can (by forward and backward reasoning) be transformed into each other. This is a weak demand; it does not imply that, for any two cases $c_1, c_2 \in C$, there exists a sequence of steps from c_1 to c_2 or from c_2 to c_1. It only demands a deducible dependency between the two cases.
4) C should be large enough such that (i) for each event $e \in E$ there is a case in C in which e has concession, and (ii) each condition $b \in B$ belongs to at least one case of C but does not belong to every case of C. This excludes self-loops and isolated conditions. We also exclude isolated events, since the occurrence of an event should be observable.

Further, we shall not allow two conditions b_1 and b_2 to have the same pre- and postset, since otherwise in every situation either both would hold or neither of them holds (or they would never be able to change). Hence two conditions are indistinguishable in the context represented in the net; they are representatives of the same condition. It is sufficient to include every condition only once in a net.

A similar argument is applicable to events with equal pre- and postsets. Any two such events either both have concession in a case or neither has concession, and the occurrence of either of them leads to the same follower case. If in a given context all important aspects of a system are represented, the significance of an event is uniquely determined by its pre- and postset.

We summarize these requirements in the following definition:

(a) Definition. A quadruple $\Sigma = (B, E; F, C)$ is called a *condition/event-system* (*C/E-system*, for short) iff

(i) $(B, E; F)$ is a simple net without isolated elements, $B \cup E \neq \emptyset$.
(ii) $C \subseteq \mathcal{P}(B)$ is an equivalence class of the *reachability relation* $R_\Sigma = (r_\Sigma \cup r_\Sigma^{-1})^*$, where $r_\Sigma \subseteq \mathcal{P}(B) \times \mathcal{P}(B)$ is given by $c_1 \, r_\Sigma \, c_2 \Leftrightarrow \exists G \subseteq E: c_1 [G\rangle \, c_2$. C is called the *case class* of Σ.
(iii) $\forall e \subseteq E \; \exists c \in C$ such that e has concession in c.

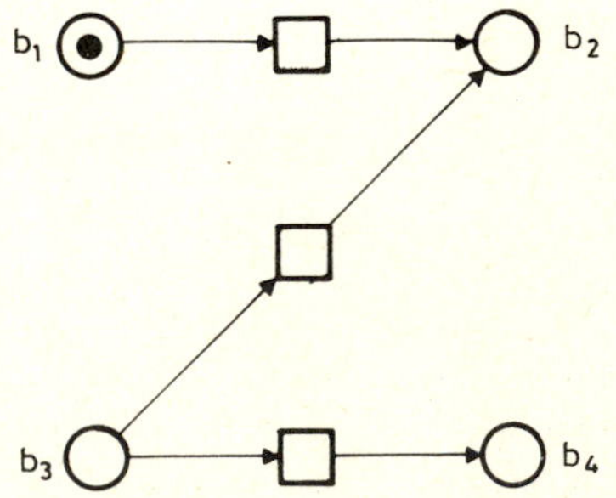

Fig. 23. A *C/E*-system, its case class is $C = \{\{b_1\}, \{b_2\}, \{b_3\}, \{b_4\}\}$

Obviously, the case class C of a C/E-system Σ is fully determined by an arbitrary element of C.

(b) Notation. Let $\Sigma = (B, E; F, C)$ be a C/E-system. Analogously to 1.5 (b) we denote B, E, F and C by B_Σ, E_Σ, F_Σ and C_Σ, respectively. Where no confusion arises we write Σ both for $B \cup E$ and for the net $(B, E; F)$.

(c) Proposition. *Let Σ be a C/E-system.*

(i) $B_\Sigma \neq \emptyset \wedge E_\Sigma \neq \emptyset \wedge F_\Sigma \neq \emptyset$.

(ii) *For $c \in C_\Sigma$, $c' \subseteq B_\Sigma$ and $G \subseteq E_\Sigma$:*
$c\,[G\rangle\, c' \Rightarrow c' \in C_\Sigma$ *and*
$c'\,[G\rangle\, c \Rightarrow c' \in C_\Sigma$.

(iii) $\forall b \in B_\Sigma\ \exists c, c' \in C_\Sigma$ *with* $b \in c \wedge b \notin c'$.

(iv) *Σ is pure.*

Proof. (i) Since $B_\Sigma \cup E_\Sigma \neq \emptyset$ and isolated elements are excluded, there exist some elements $x, y \in \Sigma$ with $x\, F_\Sigma\, y$.

(ii) follows from Definition 2.2 (a) (ii).

(iii) Since b is not isolated (2.2 (a) (i)), there is an event e in ${}^\bullet b \cup b^\bullet$. Since cases $c, c' \in C_\Sigma$ with $c\,[e\rangle\, c'$ exist and since $b \in c \cup c'$, the result follows.

(iv) An event contained in a self-loop never has concession. □

(d) Proposition. *Let Σ be a C/E-system and let $\hat{r} \subseteq \mathscr{P}(B_\Sigma) \times \mathscr{P}(B_\Sigma)$ be defined by $c_1\, \hat{r}\, c_2 \Leftrightarrow \exists\, e \in E_\Sigma : c_1\,[e\rangle\, c_2$. If E_Σ is finite then $R_\Sigma = (\hat{r} \cup \hat{r}^{-1})^*$.*

Proof. For $\hat{R} = (\hat{r} \cup \hat{r}^{-1})^*$, $\hat{R} \subseteq R_\Sigma$ trivially holds. Since with E_Σ finite every step of Σ is finite, it follows from Lemma 2.1 (e) that $r_\Sigma \subseteq \hat{r}^*$ and $r_\Sigma^{-1} \subseteq (\hat{r}^{-1})^*$. The result follows using A7 (iii) and (iv). □

2.3 Cyclic and Live Systems

The requirements for the case class C_Σ of a C/E-system Σ might not be immediately obvious; rather, one may perhaps expect C_Σ to be the set of all successor cases of some initial case. If all cases of Σ are reproducible, any such case class is identical to C_Σ.

(a) Definition. A C/E-system Σ is called *cyclic* iff $\forall c_1, c_2 \in C_\Sigma : c_1\,(r_\Sigma^*)\, c_2$.

(b) Proposition. *Let Σ be a cyclic C/E-system and let $c \in C_\Sigma$. Then $C_\Sigma = \{c' \mid c\, r_\Sigma^*\, c'\}$.*

Proof. Since Σ is cyclic $r_\Sigma^{-1} \subseteq r_\Sigma^*$. Then applying A7 (iv) $R_\Sigma \subseteq r_\Sigma^*$. □

Figures 1, 2, 20, 21, 22 show cyclic C/E-systems.

In a cyclic system every event can always reoccur.

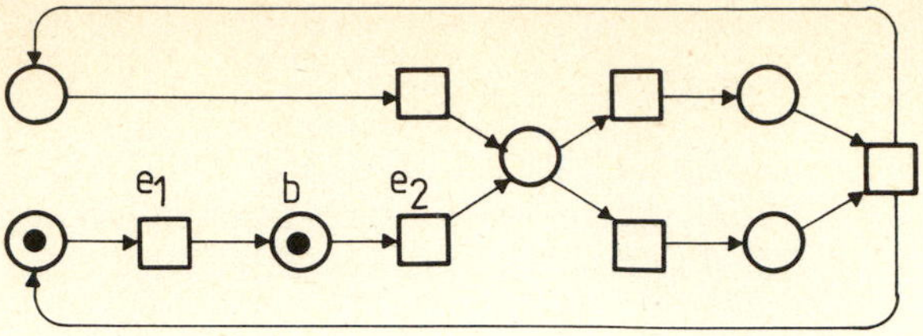

Fig. 24. A system which is live but not cyclic

(c) Definition. A C/E-system Σ is called *live* iff $\forall c \in C_\Sigma$ $\forall e \in E_\Sigma$ $\exists c' \in C_\Sigma$ such that $c\, r_\Sigma^* \, c'$ and e is c'-enabled.

(d) Proposition. *Every cyclic C/E-system is live.*

Proof. Let $c \in C_\Sigma$, $e \in E_\Sigma$. By 2.2 (a) there exists $c' \in C_\Sigma$ such that e has concession in c', and by 2.3 (a), $c\, r_\Sigma^* \, c'$. □

Figure 24 shows that not every live system is cyclic: The indicated case can not be reproduced by event occurrences.

2.4 Equivalence

The systems shown in Fig. 1 and Fig. 2 behave quite similarly: In both of them the continual change of cases yields the cyclic alternation of the four seasons. We call two C/E-systems *equivalent* if their cases and steps correspond to each other in the following way:

(a) Definition. Let Σ and Σ' be C/E-systems.

(i) Given bijections $\gamma\colon C_\Sigma \to C_{\Sigma'}$ and $\varepsilon\colon E_\Sigma \to E_{\Sigma'}$, we call Σ and Σ' *(γ, ε)-equivalent* iff for all cases $c_1, c_2 \in C_\Sigma$ and all sets of events $G \subseteq E_\Sigma$: $c_1 [G\rangle c_2 \Leftrightarrow \gamma(c_1) [\varepsilon(G)\rangle \gamma(c_2)$. (Let $\varepsilon(G) = \{\varepsilon(e) \mid e \in G\}$, cf. A9 (iii).) Σ and Σ' are called *equivalent* iff they are (γ, ε)-equivalent for some tuple (γ, ε) of bijections.

(ii) Σ and Σ' are called *isomorphic* iff the nets $(B_\Sigma, E_\Sigma; F_\Sigma)$ and $(B_{\Sigma'}, E_{\Sigma'}; F_{\Sigma'})$ are β-isomorphic for some bijection β and if $c \in C_\Sigma \Leftrightarrow \{\beta(b) \mid b \in c\} \in C_{\Sigma'}$.

(b) Notation. $\Sigma \sim \Sigma'$ iff the C/E-systems Σ and Σ' are equivalent.

(c) Proposition. *$\sim$ is an equivalence relation.*

(d) Proposition. *Equivalent C/E-systems always have the same number of cases, events and steps. They may however have a different number of conditions.*

It is obvious that the systems shown in Fig. 1 and Fig. 2 are equivalent; both are also equivalent to the system shown in Fig. 25.

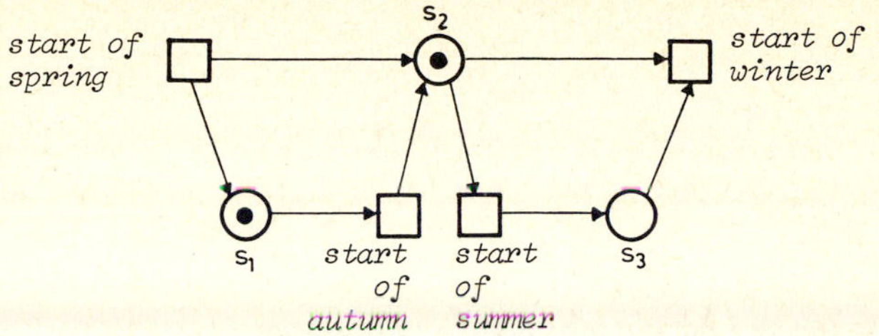

Fig. 25. A C/E-System which is equivalent to those shown in Fig. 1 and Fig. 2. Its cases are $\{s_1, s_2\} \equiv$ spring, $\{s_1, s_3\} \equiv$ summer, $\{s_2, s_3\} \equiv$ autumn, $\emptyset \equiv$ winter

(e) Proposition. *Let Σ and Σ' be two equivalent C/E-systems.*
(i) Σ *is cyclic* $\Leftrightarrow$ Σ' *is cyclic.*
(ii) Σ *is live* $\Leftrightarrow$ Σ' *is live.*

Sequential C/E-systems with single element cases (for example the system shown in Fig. 1) correspond to finite automata. For any two such systems the notion of equivalence is not interesting: it coincides with isomorphism.

(f) Lemma. *Let Σ and Σ' be C/E-systems with $\forall c \in C_\Sigma \cup C_{\Sigma'}: |c| = 1$.*
Σ and Σ' are equivalent if and only if they are isomorphic.

Proof. Let Σ be γ–ε-equivalent to Σ'. Since every case contains exactly one element, every condition b forms a case $\{b\}$ (every condition must hold in some case by Proposition 2.2 (c) (iii)). Hence γ: $C_\Sigma \to C_{\Sigma'}$ induces a bijection β': $B_\Sigma \to B_{\Sigma'}$ by means of $\beta'(b) = b' \Leftrightarrow \gamma(\{b\}) = \{b'\}$.

β: $\Sigma \to \Sigma'$, defined as $\beta(x) = \beta'(x)$ for $x \in B_\Sigma$ and $\beta(x) = \varepsilon(x)$ for $x \in E_\Sigma$, is also bijective.

Since events must be able to occur, $|{}^\bullet e| = |e^\bullet| = 1$ for all $e \in E_\Sigma$. Let $b\, F_\Sigma\, e$. Then e is $\{b\}$-enabled, therefore $\varepsilon(e)$ is $\beta(b)$-enabled and $\beta(b)\, F_{\Sigma'}\, \varepsilon(e)$. Analogously $\varepsilon(e)\, F_{\Sigma'}\, \beta(b)$ follows from $e\, F_\Sigma\, b$. The converse is trivial. □

2.5 Contact-Free C/E-Systems

In Sect. 2.1, we argued that events should not have concession in contact situations. We will now show that such situations are avoidable by means of equivalent transformations of C/E-systems. To do this, we add to each condition b its *complement* $\hat{b}$, such that in every case either b or $\hat{b}$ holds.

(a) Definition. Let Σ be a C/E-system and let $b, b' \in B_\Sigma$.
(i) b' is called the *complement* of b iff ${}^\bullet b = b'^\bullet$ and $b^\bullet = {}^\bullet b'$.
(ii) Σ is called *complete* iff each condition $b \in B_\Sigma$ has a complement $b' \in B_\Sigma$.

(b) Lemma. *Let Σ be a C/E-system and let $b \in B_\Sigma$.*
(i) *b has at most one complement. It will be denoted by $\hat{b}$.*
If b has a complement $\hat{b}$ then

(ii) $\hat{b}$ *has a complement and* $\hat{\hat{b}} = b$.
(iii) $\forall c \in C_\Sigma : b \in c \vee \hat{b} \in c$.
If Σ *is complete, then*
(iv) $\forall e \in E_\Sigma : |{}^\bullet e| = |e^\bullet|$.
(v) $\forall c \in C_\Sigma : |c| = \frac{1}{2} \cdot |B_\Sigma|$.

Proof. (i) holds since Σ is simple.
(ii) follows using the definition of a complement.
(iii) is mandatory as otherwise the involved events are in no case enabled, which contradicts Definition 2.2.
(iv) follows using the definition of a complement, since $b \in {}^\bullet e \Leftrightarrow \hat{b} \in e^\bullet$.
(v) is implied by (iii). □

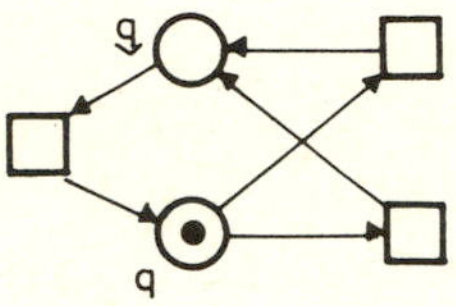

Fig. 26. A condition b and it complement $\hat{b}$

(c) Definition. Let Σ be a C/E-system and let $B \subseteq B_\Sigma$ be the set of those conditions which have no complement in B_Σ. For each $b \in B$ let $\hat{b}$ denote a new element. Let $F = \{(e, \tilde{b}) \mid (b, e) \in F_\Sigma \wedge b \in B\} \cup \{(\tilde{b}, e) \mid (e, b) \in F_\Sigma \wedge b \in B\}$.

For $c \in C_\Sigma$ let $\varphi(c) = c \cup \{\tilde{b} \mid b \in B \wedge b \notin c\}$. Then the C/E-system $\hat{\Sigma} = (B_\Sigma \cup \{\hat{b} \mid b \in B\}, E_\Sigma; F_\Sigma \cup F, \varphi(C_\Sigma))$ is the *complementation of* Σ. $\varphi(c)$ is the *complementation* of c.

Obviously, each condition b which has no complement in Σ has got $\tilde{b}$ as a complement in $\hat{\Sigma}$.

(d) Proposition. *Let* Σ *be a* C/E*-system and let* $c \in C_\Sigma$.
(i) $\hat{\hat{\Sigma}} = \hat{\Sigma}$
(ii) $\forall b \in B_\Sigma \, \forall c \in C_\Sigma : b \in \varphi(c) \Leftrightarrow \hat{b} \notin \varphi(c)$
(iii) $c = \varphi(c) \cap B_\Sigma$.

(e) Lemma. *The function* φ: $C_\Sigma \rightarrow C_{\hat{\Sigma}}$ *as defined in* 2.5 (c) *is bijective.*

Proof. φ is surjective: if $c \in C_{\hat{\Sigma}}$, $c' = c \cap B_\Sigma \in C_\Sigma$ and $\varphi(c') = c$.
φ is injective: $\varphi(c_1) = \varphi(c_2) \Rightarrow c_1 = \varphi(c_1) \cap B_\Sigma = \varphi(c_2) \cap B_\Sigma = c_2$. □

Notation. Let Σ be a C/E-system, and let $e \in E_\Sigma$. To simplify the notation, let $-e$ and $e-$ denote the pre- and postset of e in $\hat{\Sigma}$, respectively, while ${}^\bullet e$ and $e^\bullet$ will, as usual, denote the pre- and the postset of e in Σ, respectively.

(f) Proposition. *Let* Σ *be a* C/E*-system, let* $G \subseteq E_\Sigma$ *and let* B *be the set of those conditions which have no complement in* B_Σ.
(i) ${}^-G = {}^\bullet G \cup \{\tilde{b} \mid b \in B \wedge b \in G^\bullet\}$, $G^- = G^\bullet \cup \{\tilde{b} \mid b \in B \wedge b \in {}^\bullet G\}$.
(ii) ${}^\bullet G = {}^-G \cap B_\Sigma$, $G^\bullet = G^- \cap B_\Sigma$.

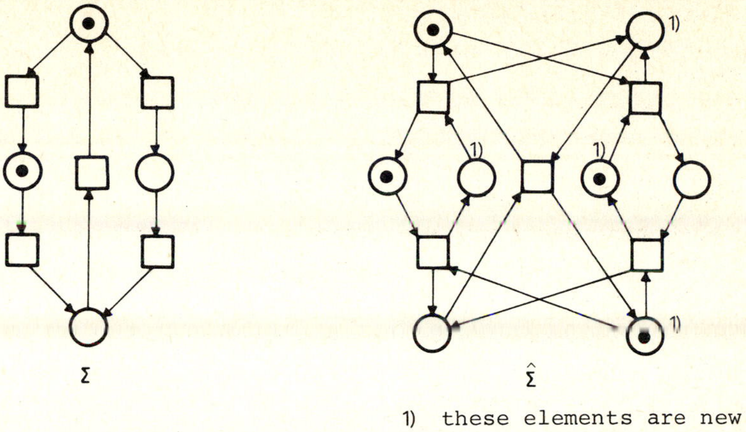

Fig. 27. A C/E-system Σ and its complementation $\hat{\Sigma}$

We are now able to show that the complementation of a C/E-system yields an equivalent contact-free system.

(g) Theorem. *If $\hat{\Sigma}$ is the complementation of a C/E-system Σ then $\hat{\Sigma}$ is equivalent to Σ.*

Proof. As $\varphi\colon C_\Sigma \to C_{\hat{\Sigma}}$ is bijective (Lemma 2.5 (e)), it is sufficient to show: $\forall c_1, c_2 \in C_\Sigma\ \forall G \subseteq R_\Sigma : c_1 [G\rangle\, c_2 \Leftrightarrow \varphi(c_1) [G\rangle\, \varphi(c_2)$. According to Lemma 2.1 (c) we show instead:

$$(c_1 \backslash c_2 = {}^\bullet G \wedge c_2 \backslash c_1 = G^\bullet) \Leftrightarrow ((\varphi(c_1) \backslash \varphi(c_2)) = {}^- G \wedge \varphi(c_2) \backslash \varphi(c_1) = G^-).$$

According to the Propositions 2.5 (d) and 2.5 (e) it holds:

$c_1 = \varphi(c_1) \cap B_\Sigma$, $c_2 = \varphi(c_2) \cap B_\Sigma$, ${}^\bullet G = {}^- G \cap B_\Sigma$ and $G^\bullet = G^- \cap B_\Sigma$, hence
$c_1 \backslash c_2 = (\varphi(c_1) \cap B_\Sigma) \backslash (\varphi(c_2) \cap B_\Sigma) = (\varphi(c_1) \backslash \varphi(c_2)) \cap B_\Sigma$ (A3 (vi))
$= {}^- G \cap B_\Sigma = {}^\bullet G.$
$c_2 \backslash c_1 = G^\bullet$ is derived in the same way.

Conversely, let B as in 2.5 (c), let $\tilde{B}_1 = \{\tilde{b} \mid b \in B \backslash c_1\}$ and let $\tilde{B}_2 = \{\tilde{b} \mid b \in B \backslash c_2\}$. Hence, $\varphi(c_1) = c_1 \cup \tilde{B}_1$ and $\varphi(c_2) = c_2 \cup \tilde{B}_2$. Now we get

$\varphi(c_1) \backslash \varphi(c_2) = (c_1 \cup \tilde{B}_1) \backslash (c_2 \cup \tilde{B}_2)$
$= (c_1 \backslash (c_2 \cup \tilde{B}_2)) \cup (\tilde{B}_1 \backslash (c_2 \cup \tilde{B}_2))$ according to A3 (iii)
$= (c_1 \backslash c_2) \cup (\tilde{B}_1 \backslash \tilde{B}_2)$, as obviously $c_1 \cap \tilde{B}_2$ and $\tilde{B}_1 \cap c_2$ are empty
$= {}^\bullet G \cup (\{\tilde{b} \mid b \in B \backslash c_1\} \backslash \{\tilde{b} \mid b \in B \backslash c_2\})$
$= {}^\bullet G \cup \{\tilde{b} \mid b \in (b \backslash c_1) \wedge b \notin (B \backslash c_2)\}$
$= {}^\bullet G \cup \{\tilde{b} \mid b \in B \wedge b \notin c_1 \wedge b \in c_2\}$
$= {}^\bullet G \cup \{\tilde{b} \mid b \in B \wedge b \in c_2 \backslash c_1\}$
$= {}^\bullet G \cup \{\tilde{b} \mid b \in B \wedge b \in G^\bullet\}$
$= {}^- G$ according to Proposition 2.5 (e).

$\varphi(c_2) \backslash \varphi(c_1) = G^-$ is derived in the same way. □

(h) Definition. Let Σ be a C/E-system.
Σ is called *contact-free* iff for each $e \in E_\Sigma$ and for each $c \in C_\Sigma$:

$$(1)\ {}^{\bullet}e \subseteq c \Rightarrow e^{\bullet} \subseteq B_\Sigma \backslash c \quad \text{and}$$
$$(2)\ e^{\bullet} \subseteq c \Rightarrow {}^{\bullet}e \subseteq B_\Sigma \backslash c.$$

Note that in (h), requirement (2) does not always follow from (1).
Example: 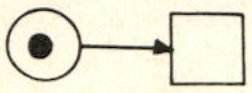

(i) Theorem. (i) *Every complete C/E-system is contact-free.*
(ii) *For every C/E-system there is an equivalent contact-free C/E-system.*
(iii) *If Σ is contact-free, then $\forall e \in E_\Sigma : {}^{\bullet}e \neq \emptyset \wedge e^{\bullet} \neq \emptyset$.*

Proof. (i) Let Σ be complete, let $b \in B_\Sigma$, $e \in E_\Sigma$ and $c \in C_\Sigma$. Then

$$b \in e^{\bullet} \cap c \Rightarrow \hat{b} \in {}^{\bullet}e \cap (B_\Sigma \backslash c) \Rightarrow {}^{\bullet}e \nsubseteq c,$$
$$b \in {}^{\bullet}e \cap c \Rightarrow \hat{b} \in e^{\bullet} \cap (B_\Sigma \backslash c) \Rightarrow e^{\bullet} \nsubseteq c.$$

(ii) $\hat{\Sigma}$ is complete (2.5 (d) (i)), contact-free ((i)) and equivalent to Σ (2.5 (g)).
(iii) Assume $e^{\bullet} = \emptyset \Rightarrow {}^{\bullet}e \neq \emptyset$ (e is not isolated). Then $\exists c \in C_\Sigma$ with ${}^{\bullet}e \subseteq c$. Since $e^{\bullet} \subseteq B_\Sigma \backslash c$, this is a contact situation. Analogously for ${}^{\bullet}e = \emptyset$. □

Of course, not every contact-free C/E-system is complete, as for example Figs. 1, 2, 20, 21, 22 show.

2.6 Case Graphs

In order to obtain an overview of all cases of a C/E-system, the construction of a case graph is useful. Its nodes are the cases and its arcs are the steps of the C/E-system.

(e) Definition. Let Σ be a C/E-system, let $\mathscr{G}$ be the set of all steps of Σ, and let $P = \{(c_1, G, c_2) \in C_\Sigma \times \mathscr{G} \times C_\Sigma \mid c_1 [G\rangle c_2\}$.

Then the graph $\Phi_\Sigma = (C_\Sigma, P)$ is called the *case graph* of Σ (for the representation of graphs see A12–A14).

(b) Theorem. *A C/E-system is cyclic if and only if its case graph is strongly connected.*

Proof. Let Σ be a C/E-system with set of steps $\mathscr{G}$. Σ is cyclic

$$\Leftrightarrow \forall c, c' \in C_\Sigma : (c\, r_\Sigma^* \, c')$$
$$\Leftrightarrow \forall c, c' \in C_\Sigma\, \exists G_1, \ldots, G_n \in \mathscr{G}\, \exists c_0, \ldots, c_n \in C_\Sigma : c_0 [G_1\rangle c_1 \ldots [G_n\rangle c_n \wedge c_0 = c \wedge c_n = c'$$
$\Leftrightarrow \Phi_\Sigma$ is strongly connected. □

(c) Theorem. *A C/E-system Σ is live if and only if for each $c_0 \in C_\Sigma$ and for each $e \in E_\Sigma$ there is a path $c_0\, l_1\, c_1 \ldots l_n\, c_n$ in Φ_Σ with $l_n = \{e\}$.*

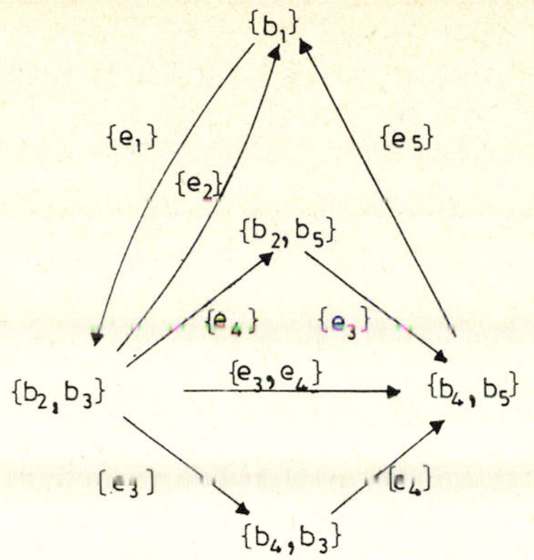

Fig. 28. The case graph corresponding to Fig. 20

Proof. Σ is live $\Leftrightarrow \forall c_0 \in C_\Sigma \; \forall e \in E_\Sigma \; \exists c, c' \in C_\Sigma : c_0 \, r_\Sigma^* \, c \wedge c\,[e\rangle\, c' \Leftrightarrow$ there is a path $c_0 \, l_1 \, c_1 \ldots c_{n-1} \, l_n \, c_n$ with $c_{n-1} = c$, $l_n = \{e\}$, $c_n = c'$. □

(d) Theorem. *Two C/E-systems are equivalent if and only if their case graphs are isomorphic.*

Proof. Let Σ and Σ' be two C/E-systems with case graphs $\Phi_\Sigma = (C_\Sigma, P)$ and $\Phi_{\Sigma'} = (C_{\Sigma'}, P')$, respectively, and let $\mathscr{G}$ be the set of steps of Σ.
Σ is $\gamma - \varepsilon$-equivalent to Σ'

$\Leftrightarrow (\forall c_1, c_2 \in C_\Sigma \; \forall G \in \mathscr{G} : c_1\,[G\rangle\, c_2 \Leftrightarrow \gamma(c_1)\,[\varepsilon(G)\rangle\, \gamma(c_2))$
$\Leftrightarrow (\forall c_1, c_2 \in C_\Sigma \; \forall G \in \mathscr{G} : (c_1, G, c_2) \in P \Leftrightarrow (\gamma(c_1), \varepsilon(G), \gamma(c_2)) \in P')$
$\Leftrightarrow \Phi_\Sigma$ is $\gamma - \varepsilon$-isomorphic to $\Phi_{\Sigma'}$. □

Not every graph can be interpreted as the case graph of a C/E-system, as shown in Fig. 29: In case c_1, e_1 and e_2 have concession. If in c_1 there is a conflict between e_1 and e_2, e_2 is not c_2-enabled, and therefore the arc $(c_2, \{e_2\}, c_4)$ is excluded. If in c_1 there is no conflict between e_1 and e_2, e_1 also has concession in c_3 and therefore the arc $(c_3, \{e_1\}, c_4)$ is required.

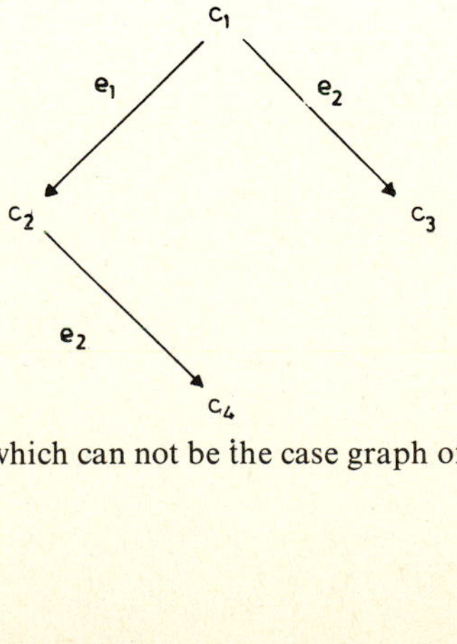

Fig. 29. A graph which can not be the case graph of any C/E-system

Case graphs quickly get very complicated in strongly concurrent systems. For example, a step consisting of n events generates 2^n arcs in the case graph.

The following theorem will be needed later:

(e) Theorem. *Let Σ be a C/E-system, let $c_1, c_2, c_3 \in C_\Sigma$ and let $G_1, G_2 \subseteq E_\Sigma$.*
(i) *If $c_1\, G_1\, c_2\, G_2\, c_3$ is a path in Φ_Σ, then $G_1 \cap G_2 = \emptyset$.*
(ii) *Let $G_1 \cap G_2 = \emptyset$. If $c_1\, (G_1 \cup G_2)\, c_3$ is an arc in Φ_Σ then there exists $c \in C_\Sigma$ such that $c_1\, G_1\, c\, G_2\, c_3$ is a path in Φ_Σ.*

Proof.
(i) $e \in G_1 \Rightarrow c_2 \cap {}^\bullet e = \emptyset \Rightarrow e$ is not c_2-enabled $\Rightarrow e \notin G_2$.
(ii) $c_1\, (G_1 \cup G_2)\, c_2$ is an arc in $\Phi_\Sigma \Rightarrow c_1\, [G_1 \cup G_2\rangle\, c_2 \Rightarrow c_1\, [G_1\rangle\, c$ and $c\, [G_2\rangle\, c_2$ where $c = (c_1 \setminus {}^\bullet G_1) \cup G_1^\bullet$. □

Exercises for Chapter 2

1. A shepherd intends to cross a river together with a goat, a wolf and a head of cabbage. With the shepherd, only one additional object fits into the boat. The situation must be avoided where a) the wolf and the goat, or b) the goat and the head of cabbage remain allone (for obvious reasons). Represent a suitable organisation for crossing.

2. Interpret the conditions s_1, s_2 and s_3 in Fig. 25.

3. Are the following C/E-systems equivalent?

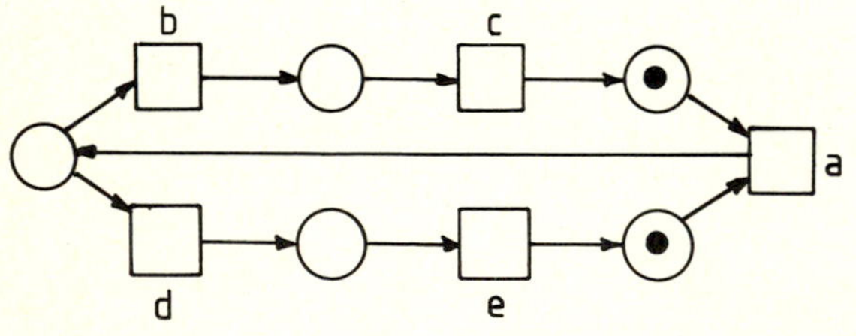

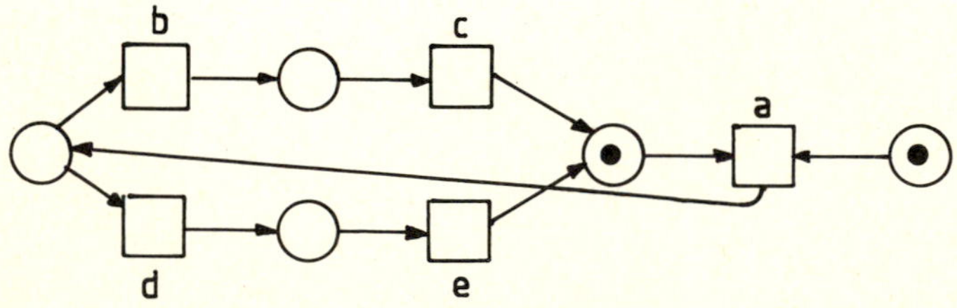

4. For the following C/E-system construct an equivalent one with a minimal number of conditions:

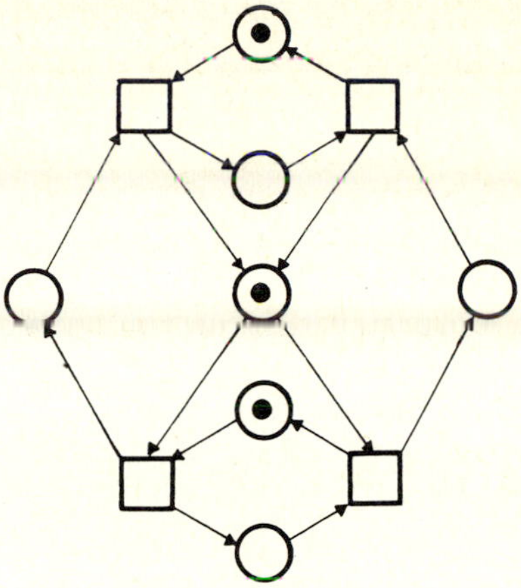

*5. Let Σ and Σ' be two C/E-systems.
   (i) Given a bijection γ: $C_\Sigma \to C_{\Sigma'}$, Σ' *γ-simulates* Σ iff $\forall G \subseteq E_\Sigma \; \exists G' \subseteq E_{\Sigma'}$ such that $c_1 [G\rangle c_2 \Rightarrow \gamma(c_1) [G'\rangle \gamma(c_2)$.
   (ii) Given a bijection ε: $E_\Sigma \to E_{\Sigma'}$, Σ' *ε-simulates* Σ iff $\forall c_1, c_2 \in C_\Sigma$ $\exists c_1', c_2' \in C_{\Sigma'}$ such that $c_1 [G\rangle c_2 \Rightarrow c_1' [\varepsilon(G)\rangle c_2'$.
   (a) Are Σ and Σ' equivalent, if Σ' γ-simulates Σ and Σ γ^{-1}-simulates Σ'?
   (b) Are Σ and Σ' equivalent, if Σ' ε-simulates Σ and Σ ε^{-1}-simulates Σ'?

6. Are the C/E-systems of the following figures contact free: Fig. 1, Fig. 2, Fig. 21, Fig. 22, Fig. 24, Fig. 25?

7. Construct the complementation of the following C/E-system:

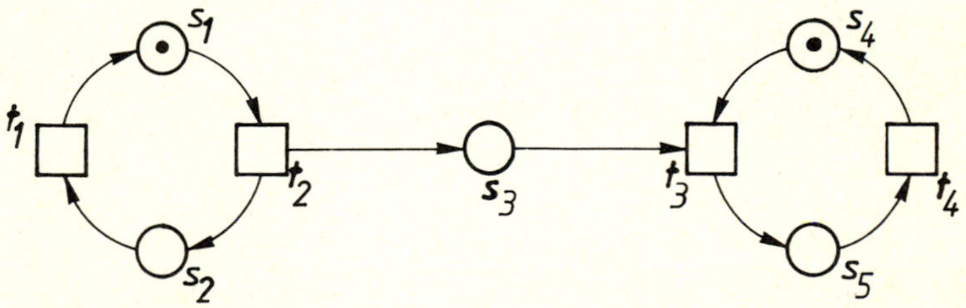

8. Construct the case graph of the C/E-system in Exercise 7.

Chapter 3

Processes of Condition/Event-Systems

This chapter deals with processes which can run on C/E-systems. One may be tempted to define a process of a C/E-system as a path of its case graph. But what we mean intuitively when speaking of processes is not adequately described by such a path: the total ordering of its elements does not give any information as to whether the events actually occur one after the other or whether they are independent of each other. The partial order in which events occur is only indirectly represented in the case graph by the set of all possibilities of occurrences as successions of steps.

We therefore search for a more convenient description of processes: one which is, in particular, unambiguous and indicates explicitly whether events occur concurrently. Such a description can be considered as a record of event occurrences and changes of conditions. The entries in this record are partially ordered by the relation "a is a causal prerequisite for b", since repetitions of the same event or the same condition are recorded as new entries. There is a fairly obvious representation of such records, namely again as a net. For instance, all of the occurrences in Fig. 20 are completely represented in Fig. 30.

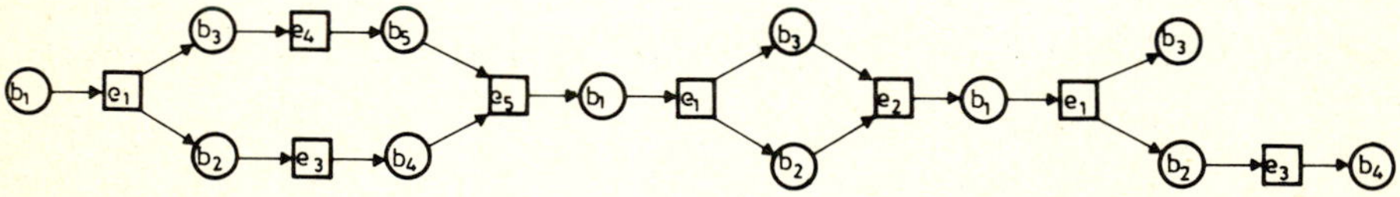

Fig. 30. A net representation corresponding to Fig. 20

A given T-element represents the occurrence of the event denoted by its labelling. Distinct T-elements with the same labelling denote several, different, occurrences of the same event. Similarly, an S-element s shows by its inscription b, that b was satisfied by the occurrence of ${}^{\bullet}s$ and ceased to hold as a result of the occurrence of $s^{\bullet}$. Just as in the corresponding concrete situations the conflicts were resolved, all S-elements are now unbranched. To facilitate the handling of such process descriptions as "partially ordered nets", we shall first study some properties of partially ordered sets and then consider *occurrence nets,* i.e. those partially ordered nets which are suitable for the description of processes. We then introduce processes and show how they can be composed and decomposed, and finally study their connection to case graphs.

3.1 Partially Ordered Sets

The relations of causal dependence and independence will turn out to be symmetric and (by definition) reflexive, but in general they will not be transitive relations. To start with, we shall consider *similarity relations:*

(a) Definition. A binary relation $\varrho \subseteq A \times A$ on a set A is called a *similarity relation* iff
(i) $\forall a \in A: a\,\varrho\, a$ (ϱ is reflexive),
(ii) $\forall a, b \in A: a\,\varrho\, b \Rightarrow b\,\varrho\, a$ (ϱ is symmetric).

A subset $B \subseteq A$ is called a *region* of a similarity relation ϱ iff
(i) $\forall a, b \in B: a\,\varrho\, b$ (ϱ is full on B),
(ii) $\forall a \in A: a \notin B \Rightarrow \exists b \in B: \neg(a\,\varrho\, b)$ (B is a maximal subset on which ϱ is full).

(b) Proposition. *Let A be b set and let $\varrho \subseteq A \times A$ be a similarity relation.*
(i) *Each element of A belongs to at least one region of ϱ.*
(ii) *Regions of a non-empty set A are not empty, and no region is a proper subset of any other region.*
(iii) *If ϱ is an equivalence relation then the regions of ϱ are exactly the equivalence classes of ϱ.*

(c) Graphical representation. A finite similarity relation over a set A can be represented uniquely as an undirected graph. A is taken as the set of nodes and $K = \{(a, b) \mid a \neq b \wedge a\,\varrho\, b)$ as the set of arcs. Figure 31 shows a similarity relation. Its regions are surrounded by broken lines.

We now consider partially ordered sets (see A11). The relations $\underline{\text{li}}$ (elements are linearly ordered, are on one line) and $\underline{\text{co}}$ (elements are unordered, are "concurrent") are defined as follows:

(d) Definition. Let A be a partially ordered set.
(i) Let $\underline{\text{li}} \subseteq A \times A$ be given by $a\ \underline{\text{li}}\ b \Leftrightarrow a < b \vee b < a \vee a = b$.
(ii) Let $\underline{\text{co}} \subseteq A \times A$ be given by $a\ \underline{\text{co}}\ b \Leftrightarrow \neg(a\ \underline{\text{li}}\ b) \vee a = b$.
(i.e. $a\ \underline{\text{co}}\ b \Leftrightarrow \neg(a < b \vee b < a)$).

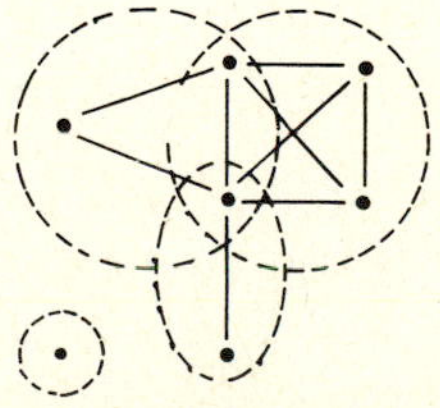

Fig. 31. A similarity relation with 4 regions

(e) Proposition. *Let A be a partially ordered set, and let* , $b \in A$.
(i) $a \ \underline{\text{li}}\ b \vee a \ \underline{\text{co}}\ b$,
(ii) $(a \ \underline{\text{li}}\ b \wedge a \ \underline{\text{co}}\ b) \Leftrightarrow a = b$.

(f) Theorem. *For any partially ordered set A, $\underline{\text{li}}$ and $\underline{\text{co}}$ are similarity relations.*

Proof. Reflexivity and symmetry of $\underline{\text{li}}$ follow immediately from the definition. The complement $A \times A \setminus \varrho$ of a symmetric relation $\varrho \subseteq A \times A$ is symmetric. The complement of $\underline{\text{li}}$ is therefore symmetric, and becomes reflexive by adding the pairs (x, x). □

Figure 32 shows a partially ordered set and the corresponding relations $\underline{\text{li}}$ and $\underline{\text{co}}$ (the graphical representation of partial orders is explained in A11).

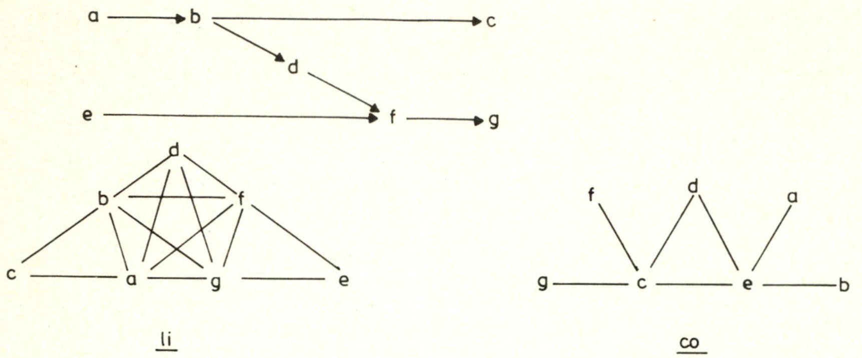

Fig. 32. A partially ordered set with its relations $\underline{li}$ and $\underline{co}$

(g) Definition. Let A be a partially ordered set, and let $B \subseteq A$.
(i) B is called a *line* iff B is a region of $\underline{\text{li}}$.
(ii) B is called a *cut* iff B is a region of $\underline{\text{co}}$.

The partial ordering in Fig. 32 yields the three lines $\{a, b, c\}$, $\{e, f, g\}$ and $\{a, b, d, f, g\}$, and the five cuts $\{e, a\}$, $\{e, b\}$, $\{e, d, c\}$, $\{f, c\}$ and $\{g, c\}$.

(h) Proposition. *Let A be a partially ordered set, and let $B \subseteq A$.*
(i) *B is a line iff*
 (a) $\forall a, b \in B : a < b \vee b < a \vee a = b$ *and*
 (b) $\forall a \in A \setminus B \ \exists b \in B$ *with* $\neg (a < b \vee b < a)$.
(ii) *B is a cut iff*
 (a) $\forall a, b \in B : \neg (a < b \vee b < a)$ *and*
 (b) $\forall a \in A \setminus B \ \exists b \in B$ *with* $a < b \vee b < a$.

(i) Definition. Let A be a partially ordered set, let $B, C \subseteq A$.
(i) A is called *bounded* iff there exists an $n \in \mathbb{N}$ such that for each line L of A, $|L| \leqq n$.
(ii) B *precedes* C (we write $B \leq C$) iff $\forall b \in B \ \forall c \in C : b < c \vee b \ \underline{\text{co}}\ c$.
 ($B < C$ means $B \leq C$ and $B \neq C$.)

(iii) Let $B^- = \{a \in A \mid \{a\} \leq B\}$ and $B^+ = \{a \in A \mid B \leq \{a\}\}$.
(iv) Let ${}^\circ B = \{b \in B \mid \forall b' \in B : b \, \underline{\text{co}} \, b' \vee b < b'\}$,
$B^\circ = \{b \in B \mid \forall b' \in B : b \, \underline{\text{co}} \, b' \vee b' < b\}$.

In particular, ${}^\circ A$ consists of the "minimal elements" of A, and A° consists of the "maximal elements" of A.

(j) Theorem. *If A is a partially ordered bounded set then ${}^\circ A$ and A° are cuts.*

Proof. Let a and b be arbitrary elements of ${}^\circ A$. Then $a \, \underline{\text{co}} \, b$ since $\neg(a < b \vee b < a)$. Let $c \in A \setminus {}^\circ A$ and let L be a line with $c \in L$. Since L is finite, there exists $d \in L \cap {}^\circ A$ and therefore $d < c$. By *Proposition* 3.1 (h) it follows that ${}^\circ A$ is a cut. Similarly it can be shown that A° is a cut. □

A line and a cut have at most one element in common:

(k) Proposition. *Let A be a partially ordered set, let L be a line and let D be a cut of A. Then $|L \cap D| \leq 1$.*

Proof. Let $a, b \in L \cap D$. Then $a \, \underline{\text{li}} \, b$, as $a, b \in L$. However $a \, \underline{\text{co}} \, b$, as $a, b \in D$. Using Corollary 3.1 (e), $a = b$. □

(l) Definition. A partially ordered set A is called *K-dense* iff each line has a non-empty intersection with each cut.

The partial ordering illustrated in Fig. 32 is K-dense, as can be easily verfied. Figure 33 shows that not every partial order is K-dense.

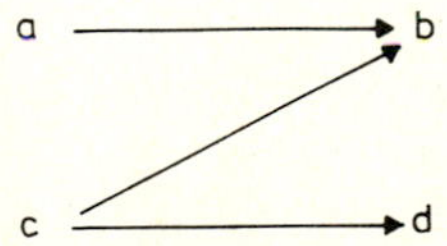

Fig. 33. A partially ordered set which is not K-dense: $\{c, b\} \cap \{a, d\} = \emptyset$

3.2 Occurrence Nets

Occurrence nets will now be introduced as cycle-free nets with unbranched S-elements. Thus, we immediately obtain a partial ordering of the elements of an occurrence net. We shall show that bounded occurrence nets are K-dense.

(a) Definition. A net $K = (S_K, T_K; F_K)$ is called an *occurrence net* if and only if
(i) $\forall a, b \in K : a (F_K^+) \, b \Leftrightarrow \neg (b \, F_K^+ \, a)$ (K is cycle-free),
(ii) $\forall s \in S_K : |{}^\bullet s| \leq 1 \wedge |s^\bullet| \leq 1$ (S-elements are unbranched).

Figure 34 shows examples of occurrence nets.

(b) Proposition. *Let K be an occurrence net. The relation <, defined by* $a < b \Leftrightarrow a F_K^+ b$, *for all* $a, b \in K$, *is a partial order on K.*

Hence, all notions concerning partially ordered sets, such as lines, cuts, boundedness and K-density are particularly defined for occurrence nets.

(c) Definition. A *slice* of an occurrence net K is a cut containing only S-elements. Let $\underline{\text{sl}}(K)$ be the set of all slices of K.

Examples of slices are shown in Fig. 34.

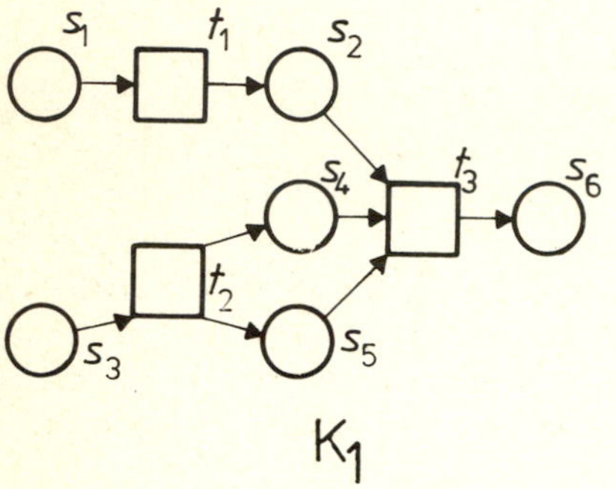

an occurrence net with three lines and 11 cuts, 5 of which are slices.
Example of a line :
$\{s_3, t_2, s_4, t_3, s_6\}$
Example of a cut :
$\{t_1, s_4, s_5\}$
a cut which is a slice :
$\{s_1, s_3\}$

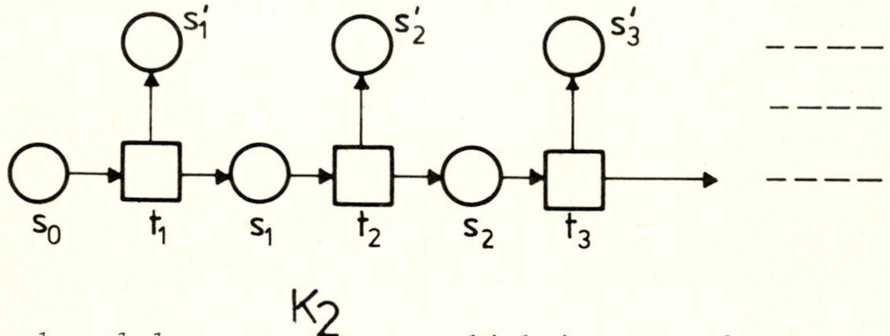

An unbounded occurrence net which is not K-dense :
$\{s_0, t_1, s_1, \ldots\} \cap \{s_1', s_2', \ldots\} = \emptyset$

Fig. 34. Examples for occurrence nets

(d) Theorem. *Every bounded non-empty occurrence net is K-dense.*

Proof. Let K be a bounded non-empty occurrence net. Assume that K is not K-dense. Let L be a line and D be a cut of K with $L \cap D = \emptyset$. Since L is not empty and finite, $x_1 = \min(L)$ and $x_2 = \max(L)$ exist. Obviously $x_1 \in {}^\circ K$ and $x_2 \in K^\circ$. Since D is a cut and $x_1 \notin D$, $\exists d \in D$ such that $x_1 < d \vee d < x_1$. As $x_1 \in {}^\circ K$, $x_1 < d$. By analogy, as $x_2 \in K^\circ$, there exists some $d' \in D$ with $d' < x_2$.

Now let $a_1 = \max\{x \in L \mid \exists d \in D \text{ with } x < d\}$ and $a_2 = \min\{x \in L \mid \exists d \in D$ with $d < x\}$. The existence of a_1 and a_2 follows now from the finiteness of L.

If $a_2 \leq a_1$, $\exists d, d' \in D$ with $d < a_2 \leq a_1 < d'$. But this is not possible since D is a cut. Therefore $a_1 < a_2$, since $a_1, a_2 \in L$. From the definition of a_1 it follows: $\exists b_1 \in a_1^{\bullet}$ $\exists d \in D$ with $b_1 \leq d$ and $\exists b_2 \in {}^{\bullet}a_2$ $\exists d' \in D$ with $d' \leq b_2$, where $b_1, b_2 \notin L$.

Since $a_1, a_2 \in L$ and $a_1 < a_2$, $\exists c_1 \in a_1^{\bullet}$ with $c_1 \in L$ and $\exists c_2 \in {}^{\bullet}a_2$ with $c_2 \in L$.

Obviously $b_1 \neq c_1$ and $b_2 \neq c_2$. Since S-elements are unbranched, $a_1, a_2 \in T_K$ follows. Therefore $(a_1, a_2) \notin F_K$. There must be at least one S-element $s \in L$ with $a_1 < s < a_2$. By definition of a_1, $\forall d \in D: s \ \underline{\text{co}}\ d$. But this is impossible since D is a cut. □

Figure 34 shows that unbounded occurrence nets are not always K-dense.

3.3 Processes

We will now define processes of C/E-systems using bounded occurrence nets. We will define this notion only for contact-free C/E-systems, the reason for this will be discussed after having given the definition. Anyway, this is no severe restriction since every C/E-system can be transformed into an equivalent contact-free system (Theorem 2.5 (i) (ii)).

Processes will be described as mappings from bounded occurrence nets into contact-free C/E-systems satisfying two requirements: (i) Each slice is mapped injectively onto a case and (ii) the mapping of a T-element to an event respects the environment of the event.

(a) Definition. Let K be a bounded occurrence net and let Σ be a contact-free C/E-system. A mapping $p\colon K \to \Sigma$ is called a *process of* Σ iff for each slice D of K and each $t \in T_K$:
(i) $p|D$ is injective $\wedge\ p(D) \in C_\Sigma$.
(ii) $p({}^{\bullet}t) = {}^{\bullet}p(t) \wedge p(t^{\bullet}) = p(t)^{\bullet}$.

In graphical representations of processes $p\colon K \to \Sigma$, every element x of K is labelled by its image $p(x)$. In this way, Fig. 30 shows a process corresponding to Fig. 20.

The property that bounded occurrence nets are K-dense is important for the use of occurrence nets to describe non-sequential processes. Every line represents a sequence of elements which are causally dependent (a sequential subprocess). A cut is interpreted as a "snapshot" of the process. One element can be seen together with different elements in different snapshots. The K-density of an occurrence net guarantees that every sequential subprocess is represented in every snapshot.

Why may this definition not be applied to arbitrary C/E-systems? It turns out that problems arise when contact enforces a certain order of event occurrences.

As an example, we consider the system shown in Fig. 24. In the represented case, e_1 may only occur after e_2, even though all preconditions of e_1 are satisfied. A process which precisely describes this sequential occurrence of e_1 and e_2 must indicate that b ceases to hold before e_1 occurs, and this cannot be achieved without introducing the complement of b as a condition of Σ.

A possibility to introduce a notion of process for arbitrary C/E-systems Σ would be to define a process of Σ to be the corresponding process of the complementation $\hat{\Sigma}$, defined as above. However, this would yield additional S-elements in processes of contact-free, but not complete systems Σ.

(b) Theorem. *For each process $p: K \to \Sigma$:*

(i) $p(S_K) \subseteq B_\Sigma \wedge p(T_K) \subseteq E_\Sigma$ *(p is sort preserving),*

(ii) $\forall x, y \in K: x\, F_K\, y \Rightarrow p(x)\, F_\Sigma\, p(y)$ *(p respects the flow relation),*

(iii) $\forall x, y \in K: p(x) = p(y) \Rightarrow x\ \underline{\mathrm{li}}\ y$ *(events and conditions are not concurrent with themselves),*

(iv) $\forall t \in T_K: {}^\bullet t \neq \emptyset \wedge t^\bullet \neq \emptyset$ *(events have prerequisites and consequences),*

(v) *for each cut D of K: $p|D$ is injective.*

Proof. (i) $p(S_K) \subseteq B_\Sigma$ follows immediately from Definition (a), as each $s \in S_K$ belongs to at least one slice. For $t \in T_K$ there exists an $x \in \Sigma$ with $x \in {}^\bullet p(t) \cup p(t)^\bullet$ (Definition 2.2 (a) (ii)). Using Definition 3.3 (a) (ii) the existence of a $y \in {}^\bullet t \cup t^\bullet$ with $p(y) = x$ follows. Since $y \in S_K$, we have $x \in B_\Sigma$ and $p(t) \in x^\bullet \cup {}^\bullet x \subseteq E_\Sigma$.

(ii) For $s \in S_K$ and $t \in T_K$: $s\, F_K\, t \Rightarrow s \in {}^\bullet t \Rightarrow p(s) \in {}^\bullet p(t) \Rightarrow p(s)\, F_\Sigma\, p(t)$. Similarly, for $t\, F_K\, s$: $s \in t^\bullet \Rightarrow p(s) \in p(t)^\bullet \Rightarrow p(t)\, F_\Sigma\, p(s)$.

(iii) For $x, y \in S_K$ the result follows immediately from the definition. For $x, y \in T_K$, $x \neq y$, $p(x) = p(y)$ implies ${}^\bullet p(x) = {}^\bullet p(y)$ and $p(x)^\bullet = p(y)^\bullet$. Now using Definition 3.3 (a) (ii) we find $p({}^\bullet x) = p({}^\bullet y)$ and $p(x^\bullet) = p(y^\bullet)$. Suppose $x\ \underline{\mathrm{co}}\ y$, then there are slices $D_1 \supseteq {}^\bullet x \cup {}^\bullet y$ and $D_2 \supseteq x^\bullet \cup y^\bullet$. Either ${}^\bullet x \cup {}^\bullet y$ or $x^\bullet \cup y^\bullet$ is non-empty, and ${}^\bullet x \cap {}^\bullet y = \emptyset = x^\bullet \cap y^\bullet$ (S-elements of K are unbranched); therefore $p|D_1$ or $p|D_2$ is not injective. Hence $x\ \underline{\mathrm{li}}\ y$.

(iv) For $t \in T_K$, using (i) we have $p(t) \in E_\Sigma$. By Theorem 2.5 (i) (iii), ${}^\bullet p(t) \neq \emptyset$ and $p(t)^\bullet \neq \emptyset$. The result follows by Definition 3.3 (a) (ii).

(v) follows from (iii) and Definition 3.3 (a) (i). □

(c) Theorem. *Let $p: K \to \Sigma$ be a process, let $T \subseteq T_K$ with $\forall t_1, t_2 \in T: t_1\ \underline{\mathrm{co}}\ t_2$. Then $\exists c_1, c_2 \in C_\Sigma$ with $c_1\, [p(T)\rangle\, c_2$.*

Proof. Obviously $\forall s_1, s_2 \in {}^\bullet T: s_1\ \underline{\mathrm{co}}\ s_2$. Then there is a slice $D \in \underline{\mathrm{sl}}(K)$ with ${}^\bullet T \subseteq D$. Definition 3.3 (a) yields $p(D) \in C_\Sigma$ and ${}^\bullet p(T) = p({}^\bullet T) \subseteq p(D)$. $\forall s \in T^\bullet\ \exists s_1 \in D$ with $s_1 < s$. Therefore $T^\bullet \cap D = \emptyset$, and also $p(D) \cap p(T^\bullet) = p(D) \cap p(T)^\bullet = \emptyset$. Hence $p(T)$ is $p(D)$-enabled, and the result follows. □

(d) Definition. Two processes $p_1: K_1 \to \Sigma$ and $p_2: K_2 \to \Sigma$ of a C/E-system Σ are called *isomorphic* iff K_1 is β-isomorphic to K_2 and $\forall x \in K_1: p_1(x) = p_2(\beta(x))$.

In the following we shall not distinguish between isomorphic processes; by "process", we shall sometimes mean either a whole equivalence class of isomorphic processes or an arbitrary representative of this equivalence class. As discussed in 1.5 (d), the elements of the underlying occurrence nets will therefore not be explicitly named in graphical representations. This convention has already been applied in Fig. 30.

Contact-free C/E-systems are fully characterized by their sets of processes: Note that a process $p: K \to \Sigma$ is actually conceived as the set of pairs $\{(x, p(x)) \mid x \in K\}$.

(e) Theorem. *Let Σ_1, Σ_2 be two contact-free C/E-systems and let P_i be the set of processes of Σ_i $(i=1,2)$. Then $P_1 = P_2 \Leftrightarrow \Sigma_1 = \Sigma_2$.*

Proof. Let $\Sigma_i = (B_i, E_i; F_i, C_i)$ $(i=1,2)$ and let $\Sigma_1 \neq \Sigma_2$. Then there exists (without loss of generality) $b \in B_1 \cup B_2$ or $e \in E_1 \cup E_2$ or $c \in C_1 \cup C_2$ such that $b \in B_1 \backslash B_2$ or $e \in E_1 \backslash E_2$ or $(b, e) \in F_1 \backslash F_2$ or $(e, b) \in F_1 \backslash F_2$ or $c \in C_1 \backslash C_2$. Then there is a step $c_1 [e'\rangle c_2$ in Σ_1 which is not possible in Σ_2 (choose $b \in c_1 \cup c_2$ or $e' = e$ or $c = c_1$ or $c = c_2$, respectively). With $K = (S, \{t\}; F)$, let $p: K \to \Sigma_1$ be a process such that $p({}^\circ K) = c_1$ and $p(K^\circ) = c_2$ and $p(t) = e'$. Then $p \in P_1 \backslash P_2$. □

3.4 The Composition of Processes

For processes p_1, p_2 we define the composition $p_1 \circ p_2$, provided that p_1 ends in the same case that p_2 starts with.

(a) Lemma. *If $p: K \to \Sigma$ is a process then ${}^\circ K$ and K° are slices of K.*

Proof. By theorem 3.1 (j), ${}^\circ K$ and K° are cuts. Since Σ is contact-free (Definition 3.3 (a)), for each $e \in E_\Sigma$, ${}^\bullet e \neq \emptyset$ and $e^\bullet \neq \emptyset$ (Theorem 2.5 (i) (iii)). ${}^\circ K \cup K^\circ \subseteq S_K$ follows from Definition 3.3 (a) (ii). □

(b) Lemma. *Let $p_i: K_i \to \Sigma$ $(i=1,2)$ be two processes with $p_1(K_1^\circ) = p_2({}^\circ K_2)$. Then there exists up to isomorphism exactly one occurrence net K, with a slice D, and a process $p: K \to \Sigma$, such that $p \mid D^- = p_1$ and $p \mid D^+ = p_2$.*

Proof. Let $K_i = (S_i, T_i; F_i)$ $(i=1,2)$ and without loss of generality $(S_1 \cup T_1) \cap (S_2 \cup T_2) = K_1^\circ = {}^\circ K_2$. $K = (S_1 \cup S_2,\ T_1 \cup T_2;\ F_1 \cup F_2)$, $D = K_1^\circ = {}^\circ K_2$, and p, defined by $p(x) = p_i(x) \Leftrightarrow x \in K_i$ $(i=1,2)$, fulfils the requirements. □

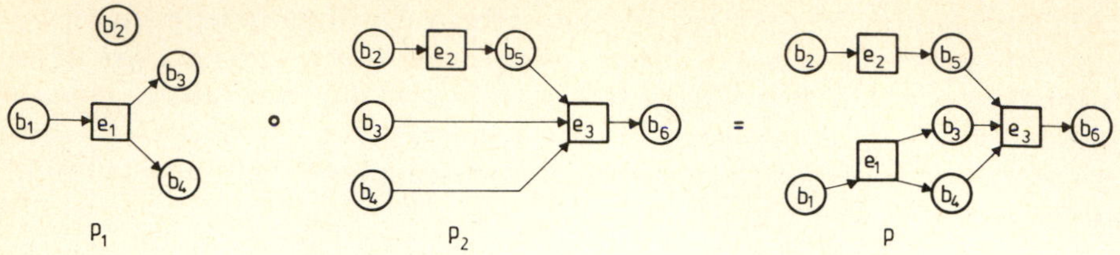

Fig. 35. Composition of processes, $p_1 \circ p_2 = p$

(c) Definition. Let p_1, p_2, p be processes, satisfying the statement of the above lemma. Then p is referred to as the *composition of* p_1 *and* p_2, and we write $p = p_1 \circ p_2$.

Each slice divides a process into composable subprocesses:

(d) Proposition. *Let* $p: K \to \Sigma$ *be a process and let* D *be a slice of* K. *Let* $p^- = p|D^-$ *and* $p^+ = p|D^+$. *Then* p^- *and* p^+ *are processes and* $p = p^- \circ p^+$.

The composition of processes is associative:

(e) Proposition. *Let* p_1, p_2, p_3 *be processes such that* $p_1 \circ p_2$ *and* $p_2 \circ p_3$ *are defined. Then* $p_1 \circ (p_2 \circ p_3)$ *and* $(p_1 \circ p_2) \circ p_3$ *are isomorphic processes.*

We call a process *elementary* if it describes a single step. Processes are decomposable into finitely many *elementary* processes.

(f) Definition. A process $p: K \to \Sigma$ is called *elementary* iff $S_K = {}^\circ K \cup K^\circ$.

As examples, the process p_1 in Fig. 35 and the processes p_3, p_4, p_5, p_6 in Fig. 36 are elementary.

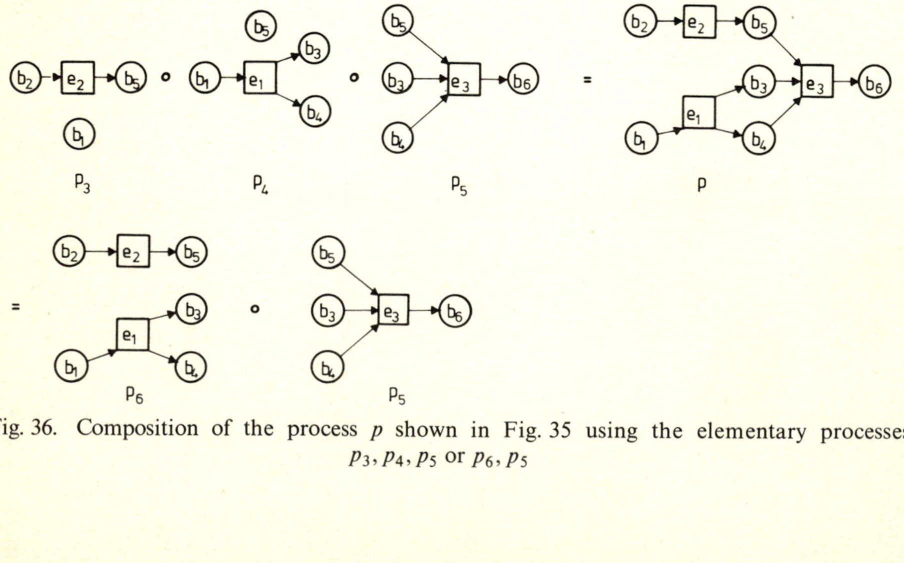

Fig. 36. Composition of the process p shown in Fig. 35 using the elementary processes p_3, p_4, p_5 or p_6, p_5

(g) Proposition. (i) *$p: K \to \Sigma$ is an elementary process iff $p({}^\circ K)\,[p(T_K)\rangle\, p(K^\circ)$ is a step of Σ.*
(ii) *If $p: K \to \Sigma$ is elementary, then for all $t_1, t_2 \in T_K$: $t_1 \underline{\text{co}}\, t_2$.*

(h) Definition. A process $p: K \to \Sigma$ is called *empty* iff $T_K = \emptyset$.

(i) Proposition. (i) *Every empty process is elementary.*
(ii) *If p' is an empty process and $p \circ p'$ (or $p' \circ p$) is defined, then $p = p \circ p'$ (or $p = p' \circ p$), respectively.*

(j) Theorem. *If $p: K \to \Sigma$ is a process then there exist finitely many elementary processes $p_1, \ldots, p_n$ such that $p = p_1 \circ \ldots \circ p_n$.*

Proof. There exists a largest number, m, of T-elements on any line of K. We prove the result by induction on m. If $m = 0$ then $T_K = \emptyset$ and p is empty. If the longest lines of K contain $m+1$ T-elements, then p is decomposable into p' and p'' such that $p = p' \circ p''$; the longest lines of p' contain m T-elements; and p'' is elementary but not empty. By the induction hypothesis, p' is composable from elementary processes $p_1, \ldots, p_n$, $p' = p_1 \circ \ldots \circ p_n$, and hence $p = p_1 \circ \ldots \circ p_n \circ p''$. □

3.5 Processes and Case Graphs

In this section we investigate the relation between processes and the paths in case graphs.

We start by showing that elementary processes directly correspond to arcs in case graphs. Then we look for paths in a case graph describing one single process. It turns out that all those paths can be transformed into each other by "decomposition" and "unification" of their arcs.

(a) Lemma. *Let Σ be a contact-free C/E-system. $p: K \to \Sigma$ is an elementary process iff there is an arc $v = (c_1, G, c_2)$ in Φ_Σ such that $p({}^\circ K) = c_1$, $p(K^\circ) = c_2$, and $p(T_K) = G$.*

Proof. If $p: K \to \Sigma$ is elementary then $p({}^\circ K)\,[p(T_K)\rangle\, p(K^\circ)$ is a step in Σ, so $(p({}^\circ K), p(T_K), p(K^\circ))$ is an arc in Φ_Σ.

Conversely, if (c_1, G, c_2) is any arc in Φ_Σ then $c_1\,[G\rangle\, c_2$. Let $K = (c_1 \cup c_2, G; F_\Sigma \cap (c_1 \cup c_2 \cup G)^2)$; then id: $K \to \Sigma$ is an elementary process of Σ. □

This lemma establishes a unique correspondence between elementary processes and arcs, and we therefore define:

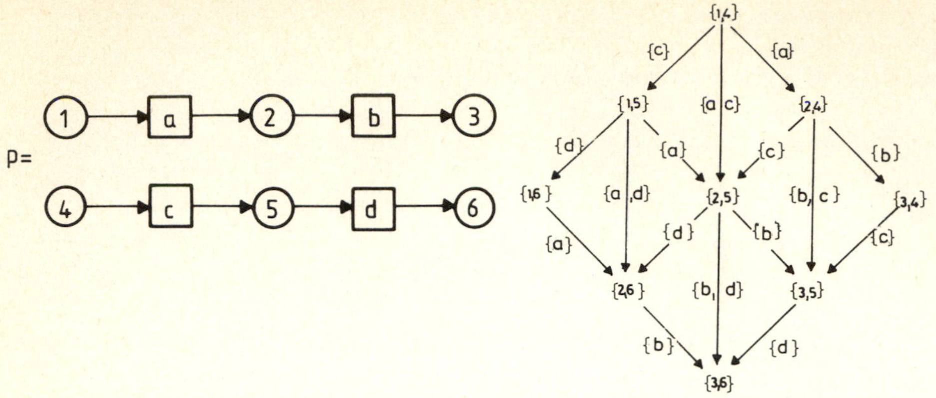

Fig. 37. A process and a part of a case graph: Each of the 13 paths from $\{1,4\}$ to $\{3,6\}$ corresponds to the process p

(b) Definition. Let Σ be a contact-free C/E-system.

(i) If v is an arc in Φ_Σ, then let $\underset{\sim}{v}$ denote the process corresponding to v, which is uniquely determined (Lemma 3.5 (a)). $\underset{\sim}{v}$ is called the *process of* v; v is called the *arc of* $\underset{\sim}{v}$.

(ii) Let $v_1, \ldots, v_n$ be arcs and let $w = v_1 \ldots v_n$ be a path in Φ_Σ. Then $\underset{\sim}{w} = \underset{\sim}{v}_1 \circ \ldots \circ \underset{\sim}{v}_n$ is called the *process of* w; w is called a *path of* $\underset{\sim}{w}$.

(iii) For $v = (c_1, G, c_2)$ and $e \in G$, let $t(v, e) = \underset{\sim}{v}^{-1}(e)$ and let $\mathscr{T}(v) = \{t(v, e) \mid e \in G\}$.

For each path of a case graph there is exactly one corresponding process. Conversely, there are in general several paths corresponding to a single process, as shown in Fig. 37. $t(v, e)$ and $\mathscr{T}(v)$ denote a single T-element and a set of T-elements of an occurrence net, respectively.

(c) Definition. Let Σ be a C/E-system, let $c_1, c_2, c_3 \in C_\Sigma$ and $G_1, G_2 \subseteq E_\Sigma$.

(i) If $u_1 = c_1 G_1 c_2$, $u_2 = c_2 G_2 c_3$ and $v = c_1 (G_1 \cup G_2) c_3$ are arcs in Φ_Σ, then the path $u_1 u_2$ is called a *decomposition* of v; v is called a *unification* of $u_1 u_2$.

(ii) Let w, w' be paths in Φ_Σ. w' is called a *permutation* of w iff there exist paths $u_1, \ldots, u_4$ such that $w = u_1 u_2 u_3$, $w' = u_1 u_4 u_3$, and u_4 is a decomposition or a unification of u_2.

(iii) Let $w_1, \ldots, w_n$ be paths in Φ_Σ. $(w_1, \ldots, w_n)$ is called a *permutation sequence* iff for $i = 1, \ldots, n-1$: w_{i+1} is a permutation of w_i.

(d) Proposition. *Let Σ be a contact-free C/E-system, let $c_1, c_2, c_3 \in C_\Sigma$, and let $G_1, G_2 \subseteq E_\Sigma$ be disjoint and nonempty.*

(i) *If $v = c_1 (G_1 \cup G_2) c_2$ is an arc in Φ_Σ then there exists a decomposition of v of the form $c_1 G_1 c G_2 c_2$, for some $c \in C_\Sigma$.*

(ii) *Let $u_1 = c_1 G_1 c_3$ and $u_2 = c_3 G_2 c_2$ be arcs of Φ_Σ, and let $\underset{\sim}{u}_1 \circ \underset{\sim}{u}_2 : K \to \Sigma$. Then $\forall t_1, t_2 \in T_K$: $t_1 \; \underline{co} \; t_2$ iff $c_1 (G_1 \cup G_2) c_2$ is an arc in ϕ_Σ.*

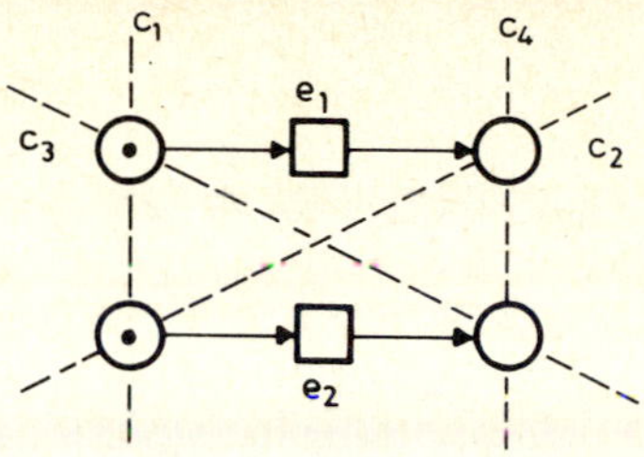

Fig. 38. $(c_1\,\{e_1\}\,c_2\,\{e_2\}\,c_4,\,c_1\,\{e_1,e_3\}\,c_4,\,c_1\,\{e_2\}\,c_3\,\{e_1\}\,c_4)$ is a permutation sequence

Proof. (i) follows immediately from Corollary 2.6 (e) (ii).

(ii) $\forall t_1, t_2 \in T_K: t_1 \,\underline{\text{co}}\, t_2$ iff there is an elementary process $p: K \to \Sigma$ with $p({}^\circ K) = c_1$, $p(K^\circ) = c_2$; and $p(T_K) = G_1 \cup G_2$ iff $c_1\,(G_1 \cup G_2)\,c_2$ is an arc in Φ_Σ (Lemma 3.5 (a)). □

(e) Lemma. *Let w be a path of some non-empty process* ($\underset{\sim}{w}: K \to \Sigma$). *Then there is a path w′ and an arc v with* $\mathscr{T}(v) = \{t \in T_K \mid {}^\bullet t \subseteq {}^\circ K\}$, *and a permutation sequence from w to v w′.*

Proof. The proof is by induction on the length, n, of w. If $n = 1$, w is an arc and the result follows immediately if we choose $v = w$ and the path w' of length 0.

If $n > 1$, there exist arcs v_1, v_2 and a path w' such that $w = w'\,v_1\,v_2$. Let $A = \{t \in \mathscr{T}(v_2) \mid {}^\bullet t \subseteq {}^\circ K\}$ and let $B = \mathscr{T}(v_2) \setminus A$ (Fig. 39).

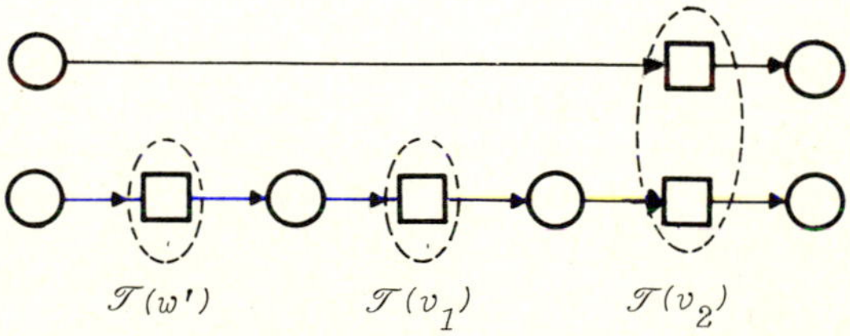

Fig. 39. Illustrating the proof of Lemma 3.5 (e)

If $A \neq \emptyset$ and $B \neq \emptyset$ then, by Proposition 3.5 (d) (i), there exists a decomposition $v_3\,v_4$ of v_2 with $\mathscr{T}(v_3) = A$ and $\mathscr{T}(v_4) = B$. Since for all $t \in A$ and for all $t' \in \mathscr{T}(v_1)$ $t \,\underline{\text{co}}\, t'$, v_1 can be unified with v_3 yielding an arc v_5 (Proposition 3.5 (d) (ii)). $w'\,v_5\,v_4$ is a permutation of w of length n. Using the induction hypothesis, $w'\,v_5$ can be permuted to a path $v'\,w''$ with $\mathscr{T}(v') = \{t \in T_K \mid {}^\bullet t \subseteq {}^\circ K\}$. $v'\,w''\,v_4$ is the required permutation.

If $B = \emptyset$, v_1 can immediately by unified with v_2. If $A = \emptyset$, the result follows from the induction hypothesis by permuting $w'\,v_1$. □

(f) Theorem. *Two paths w and w′ correspond to the same process if and only if a permutation sequence from w to w′ exists.*

Proof. Let w and w' be paths of the process $p: K \to \Sigma$. We prove the result by induction on the length n of w. $n = 1$: w is an arc. For all $t \in \mathscr{T}(w)$, ${}^\bullet t \subseteq {}^\circ K$. The permutation of w, using Lemma 3.5 (e), yields w'. Now, assume the hypothesis for paths of length $n-1$. Using Lemma 3.5 (e), we permute w and w' yielding $v w_1$ and $v' w_1'$ such that $\mathcal{T}(v) = \{t \in T_K \mid {}^\bullet t \subseteq {}^\circ K\} = \mathcal{T}(v')$. By the induction hypothesis, there exists a permutation sequence from w_1 to w_1', and the result follows, since $v = v'$.

Conversely, if $u_1 u_2$ is a decomposition of an arc v then the processes of $u_1 u_2$ and of v are equal (Proposition 3.5 (d)). Thus, if w' is a permutation of w, then w and w' are paths of the same process. Hence all elements of a permutation sequence are paths of the same process. □

Exercises for Chapter 3

1. Construct the regions of the following similarity relation:

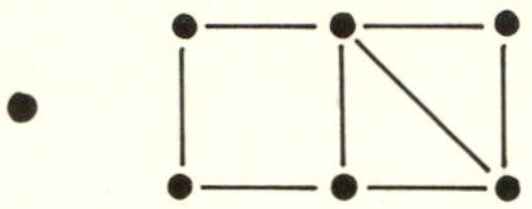

2. How many cuts, slices and lines has the following occurrence net?

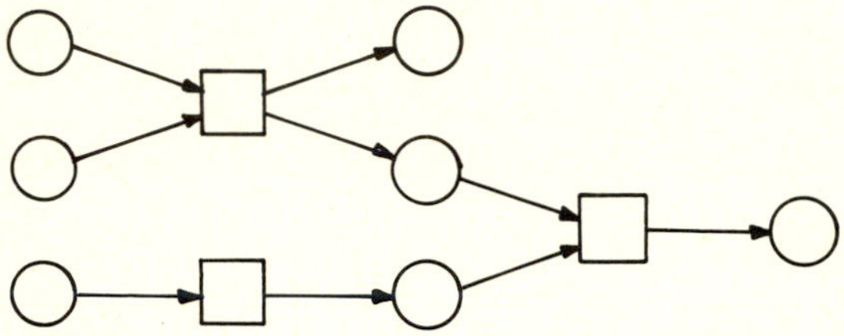

*3. Two occurrence nets K and K' are *similar* iff there exists a bijection $\tau\colon T_K \rightarrow T_{K'}$ such that $\forall t_1, t_2 \in T_K : t_1 < t_2 \Rightarrow \tau(t_1) < \tau(t_2)$.

a) For the following occurrence net construct a similar one with a minimal number of S-elements:

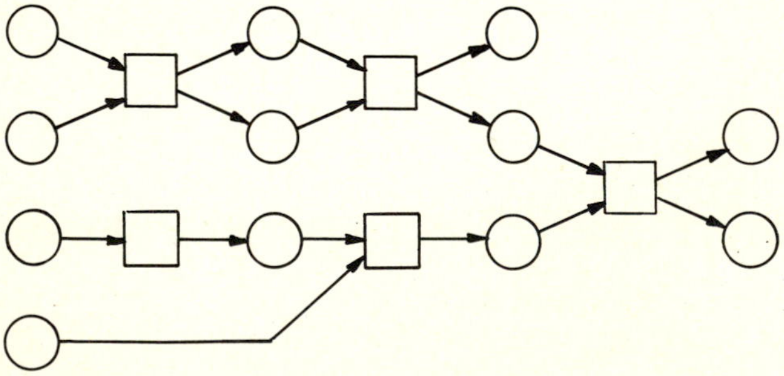

b) Let K and K' be similar occurrence nets. Does a bijection $\sigma\colon \underline{\mathrm{sl}}(K) \rightarrow \underline{\mathrm{sl}}(K')$ exist such that $\forall D_1, D_2 \in \underline{\mathrm{sl}}(K)$: $D_1 < D_2 \Rightarrow \sigma(D_1) < \sigma(D_2)$?

c) Does a bijection β exist as characterized in b), if K is finite?

4. Decompose the following process into a minimal set of elementary processes:

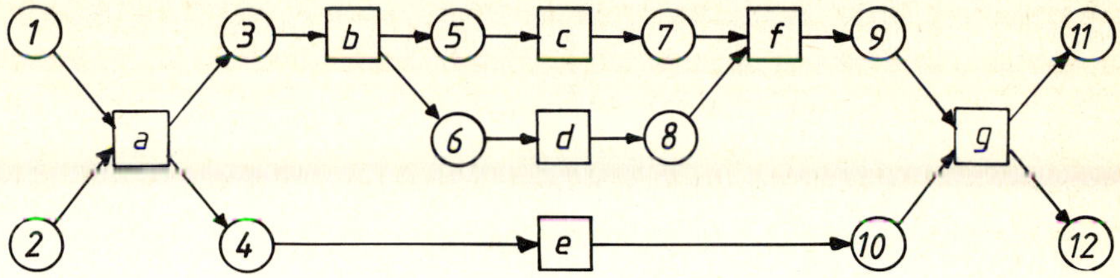

5. Construct a process of the following C/E-system:

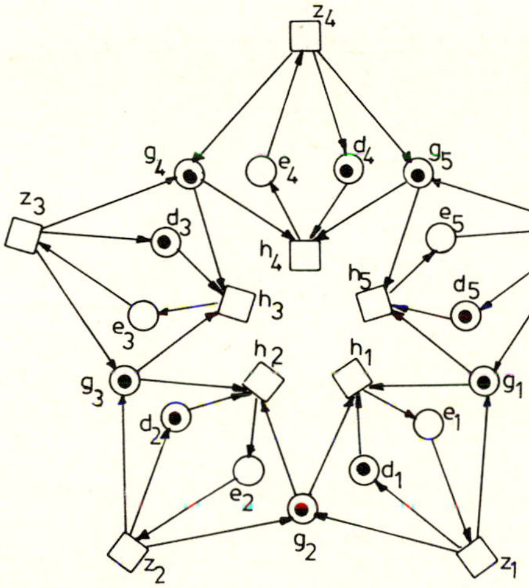

*6. Let K be a bounded occurrence net and let Σ be a C/E-system. Show that a mapping $p: K \rightarrow \Sigma$ is a process iff

(i) $p|{}^\circ K$ is injective and $p({}^\circ K) \in C_\Sigma$, and

(ii) $\forall t \in T_K: p({}^\bullet t) = {}^\bullet p(t) \wedge p(t^\bullet) = p(t)^\bullet \wedge p$ is injective on ${}^\bullet t$ and on $t^\bullet$.

Chapter 4

Properties of Systems

In the previous chapter we saw how to describe C/E-systems and how to define and analyse their dynamic behaviour. We shall now concern ourselves with some properties of C/E-systems. We shall see that some of those properties can again be described by means of the net calculus.

4.1 Synchronic Distances

An important property of a system is the degree of dependence between occurrences of its events, i.e. in which way the occurrence of a certain event is dependent on the occurrences of other events. For example, we mentioned in Chap. 1.1 (a) that the end of winter and the beginning of spring are two strongly connected (strictly "synchronized") events. Neither of them can occur without the occurrence of the other; we say that they are coincident. Events can be less tightly synchronized, for example, if their occurrences alternate (e_2 and e_3 in Fig. 22), if they are concurrent (e_1 and e_2 in Fig. 21), or if they occur in arbitrary order. At the other end of the spectrum, the occurrences of e_1 and e_2 in Fig. 22 are completely independent.

We wish to define a measure for the synchronization of events. To this end, we generalize the above considerations to pairs of sets of events, say E_1, $E_2 \subseteq E_\Sigma$. We observe how often the events of E_1 and the events of E_2, respectively, occur in each process p of the system. The absolute difference of their respective occurrence frequencies is what we call the *variance* of E_1 and E_2 in the process p. The supremum of the variances in all processes is called the *synchronic distance* $\sigma(E_1, E_2)$ of E_1 and E_2. It will turn out that σ is a metric function. Hence, synchronic distances are a means of obtaining quantitative information about the dynamic behaviour of a system without the introduction of a notion of "time".

Again we will restrict ourself to contact-free C/E-systems Σ, as the notion of synchronic distance is based on processes.

To define the synchronic distance $\sigma(E_1, E_2)$ of two sets of events, E_1, $E_2 \subseteq E_\Sigma$, we consider all processes $p: K \to \Sigma$ and count the elements of $p^{-1}(E_1)$ and $p^{-1}(E_2)$. Since we are interested in the maximal difference of the occurrences of E_1 and E_2, we count for all slices D_1, D_2 of K the elements of $p^{-1}(E_1)$ and $p^{-1}(E_2)$ between D_1 and D_2. For this, we define, for subsets M of T_K, the measure $\mu(M, D_1, D_2)$. If $D_1 < D_2$, let $\mu(M, D_1, D_2) = |M \cap D_1^{\dagger} \cap D_2^{-}|$; if

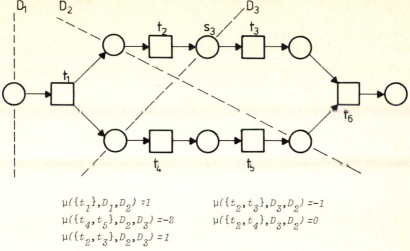

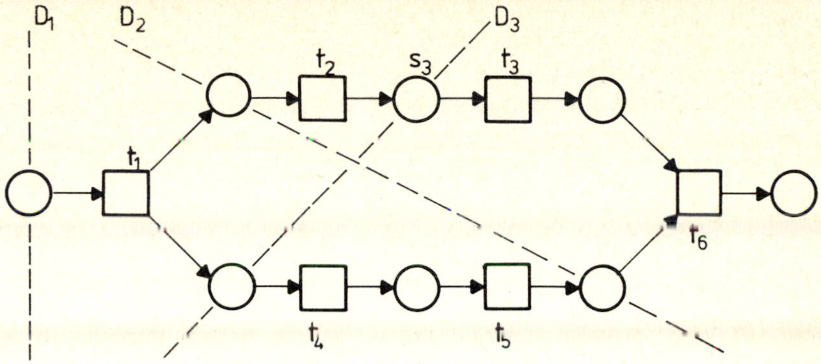

$\mu(\{t_1\}, D_1, D_2) = 1$ $\mu(\{t_2, t_3\}, D_3, D_2) = -1$
$\mu(\{t_4, t_5\}, D_2, D_3) = -2$ $\mu(\{t_2, t_4\}, D_3, D_2) = 0$
$\mu(\{t_2, t_3\}, D_2, D_3) = 1$

Fig. 40. An example for the measure μ

$D_2 < D_1$, let $\mu(M, D_1, D_2) = |M \cap D_1^- \cap D_2^+|$. However, slices may not be comparable; therefore, we define μ generally in the following way:

(a) Definition. Let K be an occurrence net, let D_1, D_2 be slices of K, and let $M \subseteq T_K$ be finite. Then let

$$\mu(M, D_1, D_2) = |M \cap D_1^+ \cap D_2^-| - |M \cap D_1^- \cap D_2^+|.$$

(b) Proposition. *For all finite subsets M of T-elements and all slices D_1, D_2 of an occurrence net K, we have $\mu(M, D_1, D_2) = -\mu(M, D_2, D_1)$.*

Using the measure μ, we now define the variance v of two sets of events in a process.

(c) Definition. Let Σ be a contact-free C/E-system. π_Σ denotes the set of all finite processes of Σ.

(d) Definition. Let Σ be a contact-free C/E-system. Let $p: K \to \Sigma \in \pi_\Sigma$ and $E_1, E_2 \subseteq E_\Sigma$.

Then $v(p, E_1, E_2) = \max\{\mu(p^{-1}(E_1), D_1, D_2) - \mu(p^{-1}(E_2), D_1, D_2) \mid D_1, D_2 \in \underline{\text{sl}}(K)\}$ is called the *variance of E_1 and E_2 in p*.

(e) Proposition. *For each process $p: K \to \Sigma$ and each pair $E_1, E_2 \subseteq E_\Sigma$: $v(p, E_1, E_2) = v(p, E_2, E_1)$.*

The synchronic distance of two sets of events can now be defined as the supremum of the variances in all finite processes.

(f) Definition. Let Σ be a contact-free C/E-system and let $E_1, E_2 \subseteq E_\Sigma$ both be finite.

$$\nu(p, \{e_0\}, \{e_3\}) = 1$$
$$\nu(p, \{e_0\}, \{e_1, e_2\}) = 2$$
$$\nu(p, \{e_0, e_1\}, \{e_2\}) = 2$$
$$\nu(p, \{e_1\}, \{e_2\}) = 2$$

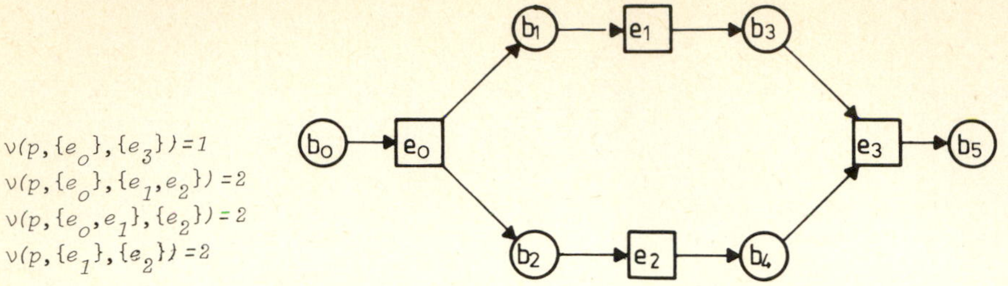

Fig. 41. Examples for the variance v

$\sigma(E_1, E_2) = \sup \{v(p, E_1, E_2) \mid p \in \pi_\Sigma\}$ is called the *synchronic distance of E_1 and E_2.*

(g) Remarks. If necessary σ is indexed to indicate the underlying C/E-system. Synchronic distances of single events are denoted by $\sigma(e_1, e_2)$ instead of $\sigma(\{e_1\}, \{e_2\})$.

(h) Graphical representation of synchronic distances. For two sets E_1, E_2 of events of a C/E-system Σ, the synchronic distance $\sigma(E_1, E_2)$ is illustrated by an additional S-element s with $\dot{}s = E_1$ and $s\dot{} = E_2$. s is not a condition of the condition/event-system Σ, but is allowed to carry arbitrarily many tokens. In each case c of Σ, s contains a number of tokens (sufficiently many tokens, in order not to hinder event occurrences). Whenever an event of E_1 or E_2 occurs, this number is increased or decreased by 1, respectively. $\sigma(E_1, E_2)$ is the supremum over the maximal variation of the number of tokens on s, yielded by finite processes. In graphical net representations, s and the new arcs are drawn as broken lines, and s is labelled by "$\sigma = x$", if $\sigma(\dot{}s, s\dot{}) = x$.

We do not provide proof here, because the newly introduced S-element s imposes a more general class of nets, which will be treated in the next chapter. In Exercise 9 of Chap. 5, we will return to this problem.

$$\sigma(\{e_4\}, \{e_0\}) = 1$$
$$\sigma(\{e_2\}, \{e_4\}) = 2$$
$$\sigma(\{e_2, e_3\}, \{e_4, e_5\}) = 4$$
$$\sigma(\{e_2, e_4\}, \{e_3, e_5\}) = 2$$
$$\sigma(\{e_4, e_5\}, \{e_3\}) = \omega$$

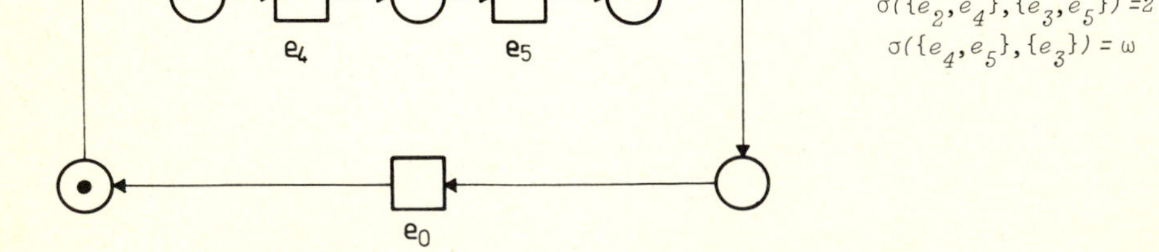

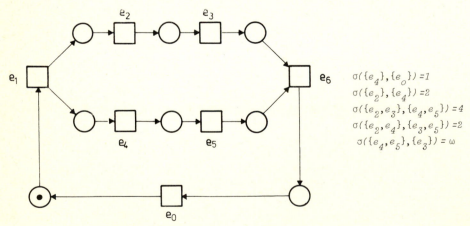

Fig. 42. Synchronic distances between sets of events

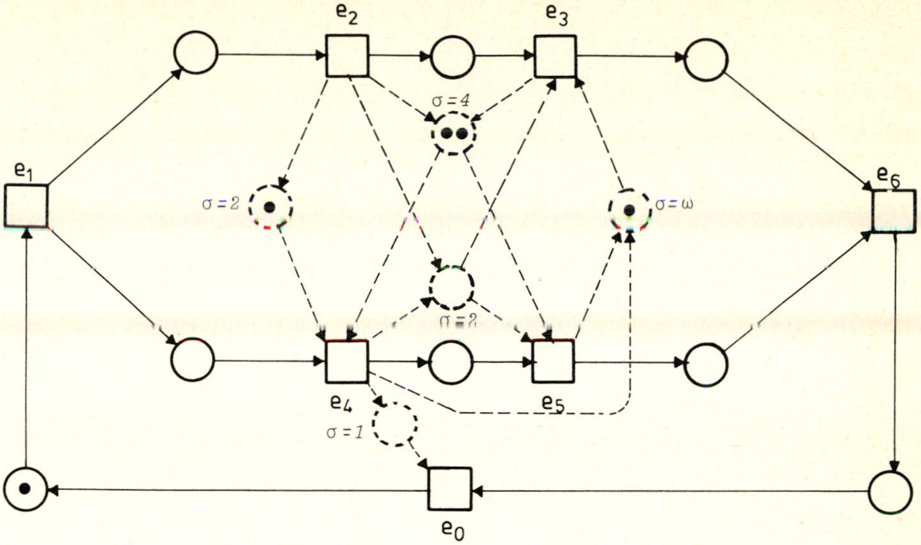

Fig. 43. Graphical representation of the synchronic distances given in Fig.42

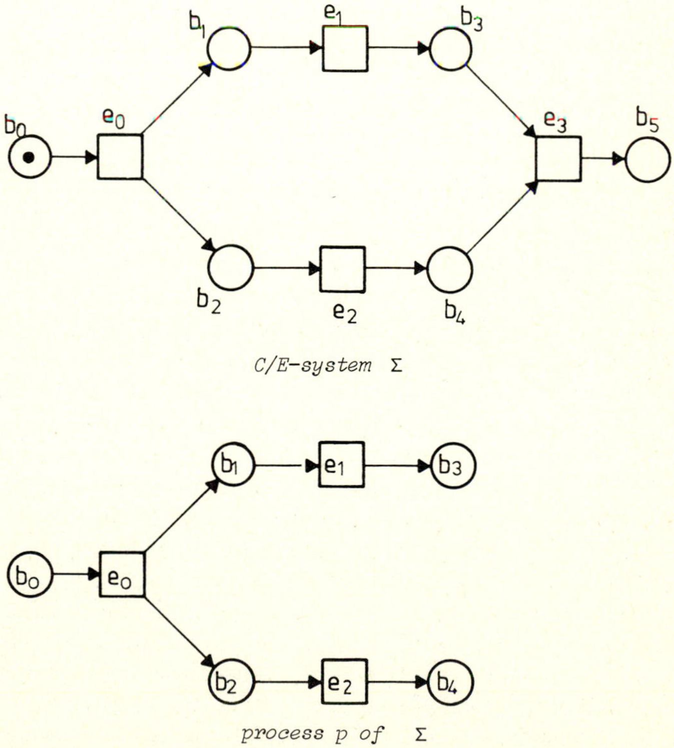

Fig. 44. A C/E-system in which the two events e_1 and e_2 occur concurrently ($\sigma(e_1, e_2) = 2$)

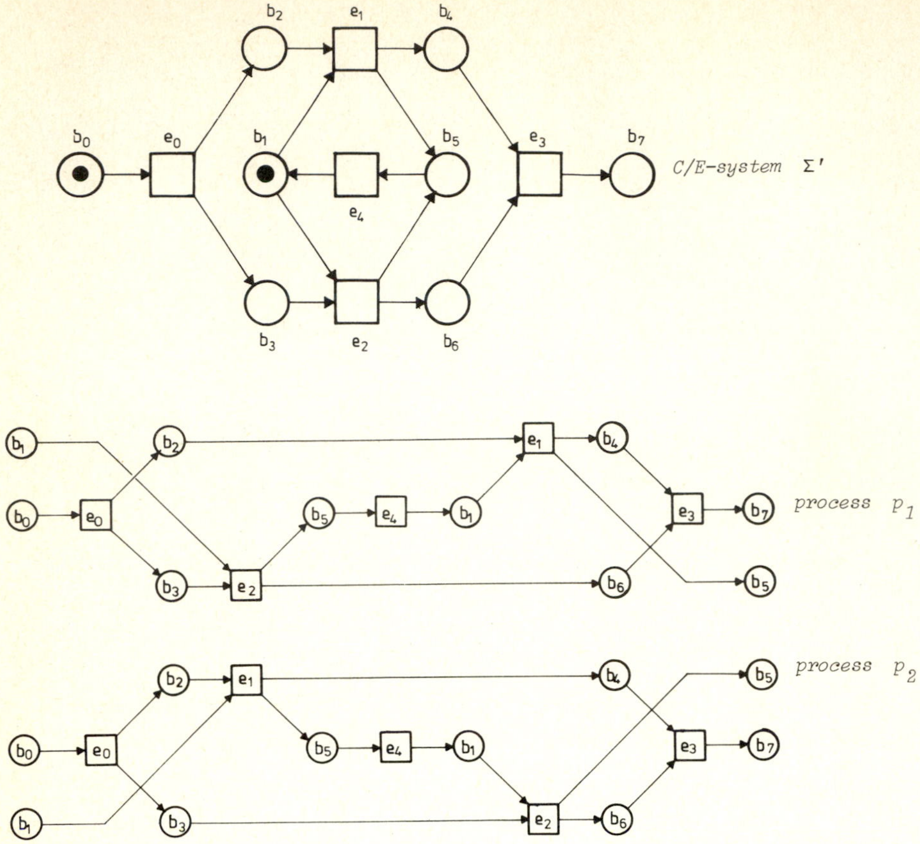

Fig. 45. A C/E-system in which e_1 and e_2 occur in some (arbitrary) order ($\sigma(e_1, e_2) = 1$)

(i) Some special synchronic distances. Obviously, we obtain a synchronic distance $\sigma(e_1, e_2) = 0$ if and only if $e_1 = e_2$; that is, e_1 and e_2 occur coincidently (as, for example, the end of winter and the beginning of spring in 1.1 (a)). Correspondingly, for sets of events E_1, E_2, $\sigma(E_1, E_2) = 0$ if and only if $E_1 = E_2$.

We now consider the two systems Σ and Σ' shown in Fig. 44 and in Fig. 45. The two events e_1, e_2 occur in Σ concurrently, they are independent. By applying the definitions, we obtain $\sigma_\Sigma(e_1, e_2) = 2$. In Fig. 45 we change the system by introducing a regulation mechanism, which prevents e_1 and e_2 from occuring concurrently, forcing them to occur in some arbitrary order. ($p_i^{-1}(e_1)$ and $p_i^{-1}(e_2)$ $(i = 1, 2)$ are situated on one line in the processes p_1 and p_2 of Σ', while $p^{-1}(e_1)$ and $p^{-1}(e_2)$ are concurrent in the process p of Σ.) The conceptual difference of the systems Σ and Σ' is reflected by the synchronic distance of e_1 and e_2. In the system Σ', we find $\sigma_{\Sigma'}(e_1, e_2) = 1$. This example shows how synchronic distances may describe the difference between concurrency ($\sigma(e_1, e_2) = 2$) and occurrence in some (possibly unspecified) order.

In Fig. 46, corresponding pairs of events of Σ_1 and Σ_2, respectively, have the same synchronic distances: $\sigma(e_1, e_2) = \sigma(e_1, e_4) = \omega$ and $\sigma(e_1, e_3) =$

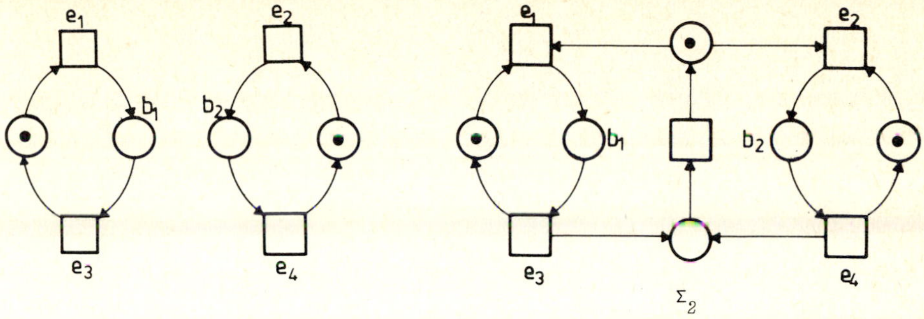

Fig. 46. Two C/E systems Σ_1, Σ_2 with $\sigma_{\Sigma_1}(e, e') = \sigma_{\Sigma_2}(e, e')$ for $e, e' \in \{e_1, \ldots, e_4\}$

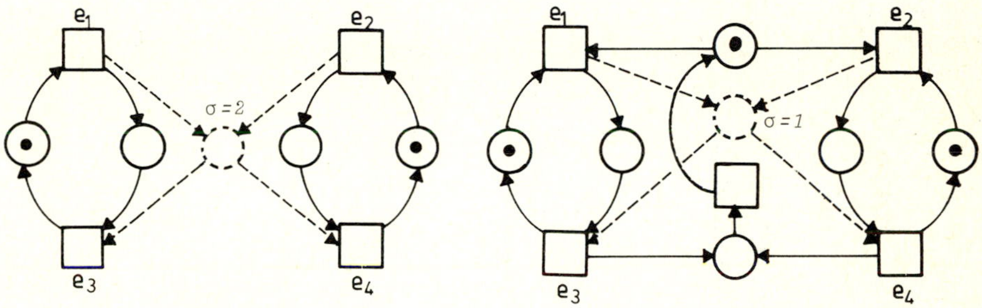

Fig. 47. Other synchronic distances in the systems of Fig. 46

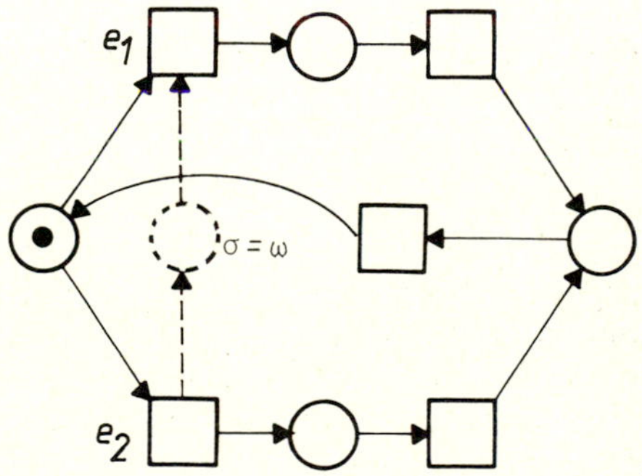

Fig. 48. An infinite synchronic distance because of a conflict

$\sigma(e_2, e_4) = 1$ in both systems. But intuitively, Σ_2 is "more strictly synchronized", as in Σ_2 no two events may occur concurrently. This is expressed by the synchronic distance of $\{e_1, e_2\}$ and $\{e_3, e_4\}$, which is 2 in Σ_1, but 1 in Σ_2 (Fig. 47).

In the system shown in Fig. 48 the events e_1 and e_2 are unboundedly often in conflict with each other; we obtain an infinite synchronic distance. In Fig. 49, the synchronic distance of e_1 and e_2 is also infinite. But in contrast to the system shown in Fig. 48, the occurrences of e_1 and e_2 are dependent on

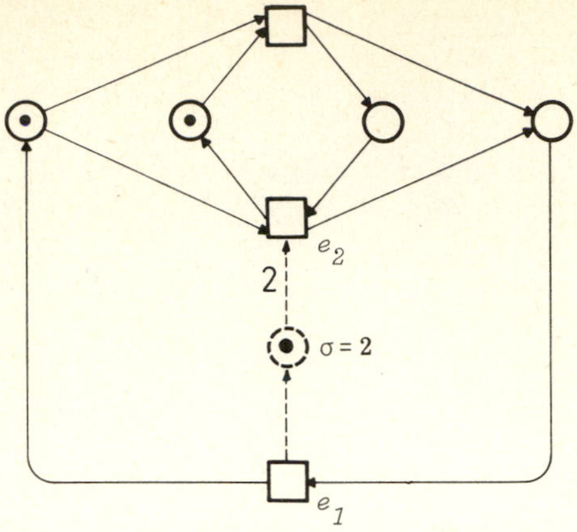

Fig. 49. A weighted synchronic distance

each other: e_1 occurs twice as often as e_2. To express this, we need to generalize the concept of a synchronic distance. In Fig. 49, for example, we specify that the occurrence of e_2 reduces the number of tokens on the new S-element by 2. In the graphical representation, the corresponding arc is labelled by the "weight" 2. This concept of *weighted synchronic distances* is not explained any further here.

4.2 Some Quantitative Properties of Synchronic Distances

First, we show that synchronic distances define a metric on the sets of events of a C/E-system. Then some other properties of synchronic distances are proved.

(a) Theorem. *Let Σ be a contact-free C/E-system, let $E_1, E_2, E_3 \subseteq E_\Sigma$. Then*
(i) $\sigma(E_1, E_2) = 0 \Leftrightarrow E_1 = E_2$,
(ii) $\sigma(E_1, E_2) = \sigma(E_2, E_1)$,
(iii) $\sigma(E_1, E_2) \leq \sigma(E_1, E_3) + \sigma(E_3, E_2)$.

Proof. (i) and (ii) follow immediately from Definition 4.1 (f). To prove (iii), let $p: K \to \Sigma \in \pi_\Sigma$ and let D_1 and D_2 be slices of K such that $v(p, E_1, E_2) = \mu(p^{-1}(E_1), D_1, D_2) - \mu(p^{-1}(E_2), D_1, D_2)$. Then, defining $[E_i] = \mu(p^{-1}(E_i), D_1, D_2)$ $(i = 1, 2, 3)$, we have: $v(p, E_1, E_2) = [E_1] - [E_2] = [E_1] - [E_3] + [E_3] - [E_2] \leq v(p, E_1, E_3) + v(p, E_3, E_2)$. Using A16 we obtain: $\sigma(E_1, E_2) = \underline{\sup}\{v(p, E_1, E_2) \mid p \in \pi_\Sigma\} \leq \underline{\sup}\{v(p, E_1, E_3) + v(p, E_3, E_2) \mid p \in \pi_\Sigma\} \leq \underline{\sup}\{v(p, E_1, E_3) \mid p \in \pi_\Sigma\} + \underline{\sup}\{v(p, E_3, E_2) \mid p \in \pi_\Sigma\}$. □

(b) Theorem. *Let Σ be a contact-free C/E-system and let $E_1, \ldots, E_4 \subseteq E_\Sigma$. Then $\sigma(E_1 \cup E_2, E_3 \cup E_4) \leq \sigma(E_1, E_3) + \sigma(E_2, E_4) + \sigma(E_1 \cap E_2, E_3 \cap E_4)$.*

Proof. Let $p\colon K \to \Sigma \in \pi_\Sigma$ and let D_1, D_2 be slices of K such that $\nu(p, E_1 \cup E_2, E_3 \cup E_4) = \mu(p^{-1}(E_1 \cup E_2), D_1, D_2) - \mu(p^{-1}(E_3 \cup E_4), D_1, D_2)$. For $E \subseteq E_\Sigma$ let $[E] = \mu(p^{-1}(E), D_1, D_2)$. Obviously for all $E, E' \subseteq E_\Sigma$: $[E \cup E'] = [E] + [E' \setminus E]$, $[E \setminus E'] = [E] - [E \cap E']$ and $[E] - [E'] \leq \nu(p, E, E') \leq \sigma(E, E')$. Therefore $\nu(p, E_1 \cup E_2, E_3 \cup E_4) = [E_1 \cup E_2] - [E_3 \cup E_4] = [E_1] + [E_2 \setminus E_1] - [E_3] - [E_4 \setminus E_3] = [E_1] + [E_2] - [E_2 \cap E_1] - [E_3] - [E_4] + [E_4 \cap E_3] \leq \nu(p, E_1, E_3) + \nu(p, E_2, E_4) + \nu(p, E_1 \cap E_2, E_3 \cap E_4)$.

The result follows using A16, as in the proof of the above theorem. □

(c) Corollary. *Let Σ be a contact-free C/E-system and let $E_1, \ldots, E_4 \subseteq E_\Sigma$ such that $E_1 \cap E_2 = \emptyset = E_3 \cap E_4$. Then $\sigma(E_1 \cup E_2, E_3 \cup E_4) \leq \sigma(E_1, E_3) + \sigma(E_2, E_4)$.*

Proof. Since $\sigma(\emptyset, \emptyset) = 0$ (Theorem 4.2 (a) (i)), the result follows immediately by application of Theorem 4.2 (b). □

(d) Theorem. *Let Σ be a contact-free C/E-system and let $E_1, E_2 \subseteq E_\Sigma$. Then $\sigma(E_1, E_2) = \sigma(E_1 \setminus E_2, E_2 \setminus E_1)$.*

Proof. Let $p\colon K \to \Sigma \in \pi_\Sigma$ and let $D_1, D_2 \in \underline{\text{sl}}(K)$. For $E \subseteq E_\Sigma$ let $[E] = \mu(p^{-1}(E), D_1, D_2)$. Then $[E_1] - [E_2] = [(E_1 \setminus E_2) \cup (E_1 \cap E_2)] - [(E_2 \setminus E_1) \cup (E_1 \cap E_2)] = [E_1 \setminus E_2] + [E_1 \cap E_2] - [E_2 \setminus E_1] - [E_1 \cap E_2] = [E_1 \setminus E_2] - [E_2 \setminus E_1]$.

Hence $\nu(p, E_1, E_2) = \nu(p, E_1 \setminus E_2, E_2 \setminus E_1)$; the result follows. □

4.3 Synchronic Distances in Sequential Systems

In purely sequential systems, synchronic distances are not very interesting. For any pair of single events we always obtain one of the values 0, 1 or ω.

(a) Definition. A *C/E*-system is called a *state machine* iff
(i) $\forall e \in E_\Sigma : |{}^\bullet e| = |e^\bullet| = 1$,
(ii) $\forall c \in C_\Sigma : |c| = 1$.

The Figs. 1 and 13 show examples of state machines.

(b) Theorem. *Let Σ be a state machine and let $e_1, e_2 \in E_\Sigma$. Then $\sigma(e_1, e_2) \in \{0, 1, \omega\}$.*

Proof. Each process of Σ consists of a line of the form

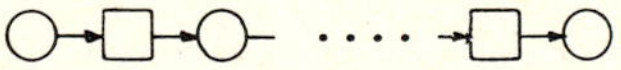

Assume, that there exists a process $p\colon K \to \Sigma$ with two T-elements $t_1, t_2 \in T_K$, such that, for $i = 1$ or $i = 2$, $p(t_1) = p(t_2) = e_i$ and $\forall t \in t_1^\dagger \cap t_2^- : p(t) \neq e_i$. Then, with $p_1 = p|({}^\bullet t_1^\dagger \cap {}^\bullet t_2^-)$, $p_n = \underbrace{p_1 \circ \ldots \circ p_1}_{n\text{-times}}$ is a process, and $\nu(p_n, \{e_1\}, \{e_2\}) \geqq n$. Then $\sigma(e_1, e_2) = \omega$.

Otherwise, for all processes p of Σ, $\nu(p, e_1, e_2) \leq 1$ and therefore $\sigma(e_1, e_2) \leq 1$. □

4.4 Synchronic Distances in Cyclic Systems

The definition of synchronic distances in 4.1 takes account of the fact that, in a process, concurrency may yield slices which are not ordered. This is important if the C/E-system is non-cyclic because the values corresponding to the situations discussed in 4.1 (i) could otherwise not be obtained. We are now going to define a simpler function σ', which is equivalent to the synchronic distance σ in the special case of cyclic C/E-systems.

(a) Definition. Let Σ be a C/E-system which is contact-free, let $E_1, E_2 \subseteq E_\Sigma$ and let $p \in \pi_\Sigma$. We define $v'(p, E_1, E_2) = \| p^{-1}(E_1) | - | p^{-1}(E_2) \|$ and $\sigma'(E_1, E_2) = \underline{\sup} \{v'(p, E_1, E_2) | p \in \pi_\Sigma\}$.

(b) Proposition. *For any arbitrary C/E-system Σ and $E_1, E_2 \subseteq E_\Sigma$: $\sigma'(E_1, E_2) \leq \sigma(E_1, E_2)$.*

For example, in Fig. 44, $\sigma'(\{e_1\}, \{e_2\}) = 1 < \sigma(\{e_1\}, \{e_2\}) = 2$.

(c) Theorem. *Let Σ be a C/E-system which is contact-free and cyclic. Then for all $E_1, E_2 \subseteq E_\Sigma$, $\sigma'(E_1, E_2) = \sigma(E_1, E_2)$.*

Proof. By Proposition 4.3 (b), it is sufficient to show $\sigma'(E_1, E_2) \geq \sigma(E_1, E_2)$. To prove this, we construct for each process p of Σ a process p' of Σ such that $v'(p', E_1, E_2) \geq v(p, E_1, E_2)$.

Let $p: K \to \Sigma$ be given. Let D_1, D_2 be slices of K with $v(p, E_1, E_2) = \mu(p^{-1}(E_1), D_1, D_2) - \mu(p^{-1}(E_2), D_1, D_2)$. Since Σ is cyclic, a process p': $K' \to \Sigma$ and a slice D_3 of K' exist such that $p \circ p'$ is a process of Σ and $p(D_3) = p(D_2)$ (see Fig. 50). Then $D_1 < D_3$ and $D_2 < D_3$.

For slices D, D' with $D < D'$, we define the process $p_{D,D'}$ by $p_{D,D'} = p | (D^+ \cap D'^-)$.

If $v'(p_{D_2, D_3}, E_1, E_2) > 0$, let $p'' = \underbrace{p_{D_2, D_3} \circ \ldots \circ p_{D_2, D_3}}_{v(p, E_1, E_2) \text{ times}}$, and we obtain $v'(p'', E_1, E_2) \geq v(p, E_1, E_2)$. Now assume

$$v'(p_{D_2, D_3}, E_1, E_2) = \| p_{D_2, D_3}^{-1}(E_1) | - | p_{D_2, D_3}^{-1}(E_2) \| = 0.$$

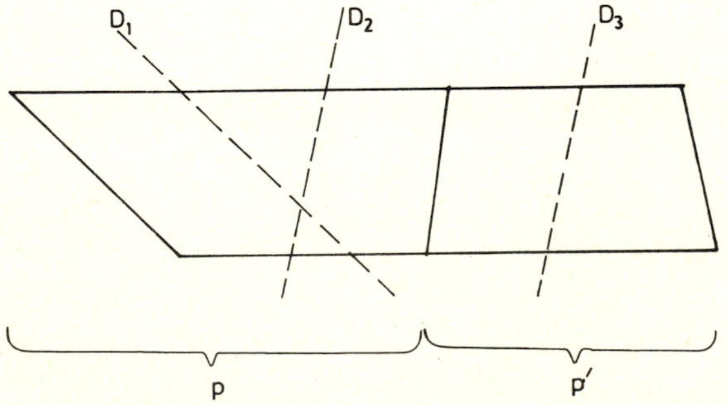

Fig. 50. Illustrating the proof of Theorem 4.4 (c)

Clearly, $|p_{D_1,D_3}^{-1}(E_i)| = |p_{D_2,D_3}^{-1}(E_i)| + |p^{-1}(E_i) \cap D_1^+ \cap D_2^-|$
$$- |p^{-1}(E_i) \cap D_1^- \cap D_2^+|.$$

Then $v'(p_{D_1,D_3}, E_1, E_2)$
$$= \| p_{D_2,D_3}^{-1}(E_1)| + |p^{-1}(E_1) \cap D_1^+ \cap D_2^-| - |p^{-1}(E_1) \cap D_1^- \cap D_2^+|$$
$$- |p_{D_2,D_3}^{-1}(E_2)| + |p^{-1}(E_2) \cap D_1^+ \cap D_2^-| - |p^{-1}(E_2) \cap D_1^- \cap D_2^+\| = 0$$
$$= |\mu(p^{-1}(E_1), D_1, D_2) - \mu(p^{-1}(E_2), D_1, D_2)|$$
$$= v(p, E_1, E_2).$$
□

4.5 Facts

It is possible to construct formulae of propositional logic by using the conditions of a C/E-system. Since conditions are allowed to change, such formulae will be true or false depending on which case the system is in. Formulae which are true in all cases of the system are especially interesting, because they describe invariant properties of the system. We shall now show how the representation and evaluation of such formulae can be integrated into the net calculus.

Consider again the C/E-system Σ_1 of Fig. 46, consisting of two simple sequential cycles. We now add the requirement that b_1 and b_2 do not hold together in any case of the system. We can achieve this by the construction of Σ_2 shown in Fig. 46. The new property of the system can be expressed in the net calculus by adding a new T-element t with $^\bullet t = \{b_1, b_2\}$ and $t^\bullet = \emptyset$, as shown in Fig. 51, which is enabled in no case of the system.

We first study the relations between formulae consisting of conditions of a C/E-system (for example $\neg(b_1 \wedge b_2)$ in Fig. 51) and the possibility of events being enabled. To this end, we consider a condition b as an atomic propositional formula, which is true in a given case c if and only if b belongs to c. Then we can construct formulae of propositional logic and evaluate their truth values.

(a) Definition. Let Σ be a C/E-system.

(i) The set A_Σ of *formulae* (of propositional logic) over B_Σ is the smallest set such that

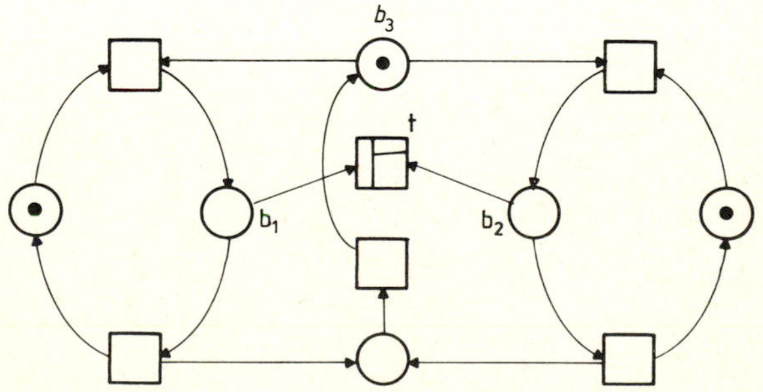

Fig. 51. Enhancement of Σ_2 of Fig. 46 by a T-element t which is never enabled

(1) $B_\Sigma \subseteq A_\Sigma$,
(2) $a_1, a_2 \in A_\Sigma \Rightarrow (a_1 \wedge a_2) \in A_\Sigma,\ (a_1 \vee a_2) \in A_\Sigma,$
$(a_1 \rightarrow a_1) \in A_\Sigma,\ (\neg a_1) \in A_\Sigma.$

(ii) Each case $c \in C_\Sigma$ induces for each $a \in A_\Sigma$ a value $\hat{c}\,(a)$, defined by

$$\begin{aligned}
\hat{c}\colon\quad & A_\Sigma \rightarrow \{0,1\} \\
& b \mapsto 1 \text{ iff } b \in c, \\
& b \mapsto 0 \text{ iff } b \notin c, \\
(a_1 \wedge a_2) & \mapsto \min(\hat{c}\,(a_1), \hat{c}\,(a_2)), \\
(a_1 \vee a_2) & \mapsto \max(\hat{c}\,(a_1), \hat{c}\,(a_2)), \\
(a_1 \rightarrow a_2) & \mapsto \hat{c}\,((\neg a_1) \vee a_2), \\
(\neg a_1) & \mapsto 1 - \hat{c}\,(a_1).
\end{aligned}$$

We interpret 1 as "true" and 0 as "false", and we call a formula a *valid in the case* c iff $\hat{c}\,(a) = 1$.

(iii) Two formulae $a_1, a_2 \in A_\Sigma$ are called *equivalent in* Σ iff for all $c \in C_\Sigma$: $\hat{c}\,(a_1) = \hat{c}\,(a_2)$.

We shall omit unnecessary brackets (note that $\wedge$ and $\vee$ are associative operators).

Next we shall associate a formula $a(e)$ with each event e of a C/E-system in such a way that for all cases c: $a(e)$ is valid in c if and only if e is not c-enabled.

(b) Definition. Let Σ be a finite C/E-system and let $e \in E_\Sigma$. Let ${}^{\bullet}e = \{b_1, \ldots, b_n\}$, $e^{\bullet} = \{b'_1, \ldots, b'_m\}$. Then $a(e)$ is the formula $(b_1 \wedge \ldots \wedge b_n) \rightarrow (b'_1 \vee \ldots \vee b'_m)$. If $e^{\bullet} = \emptyset$, then $a(e)$ is the formula $\neg(b_1 \wedge \ldots \wedge b_n)$. If ${}^{\bullet}e = \emptyset$, then $a(e)$ is the formula $b'_1 \vee \ldots \vee b'_m$.

(c) Lemma. *Let Σ be a finite C/E-system and let $e \in E_\Sigma$. Then for each $c \in C_\Sigma$, $a(e)$ is valid in c iff e is not c-enabled.*

Proof. $\hat{c}\,(a(e)) = 1 \Leftrightarrow \exists b \in {}^{\bullet}e$ with $\hat{c}\,(b) = 0$ or $\exists b' \in e^{\bullet}$ with $\hat{c}\,(b') = 1 \Leftrightarrow$ $\exists b \in {}^{\bullet}e$ with $b \notin c$ or $\exists b' \in e^{\bullet}$ with $b' \in c \Leftrightarrow e$ is not c-enabled. □

We showed above how to associate a formula to an event of a C/E-system. Next we consider how to represent arbitrary valid formulae built from conditions of the system.

For this we enlarge a C/E-system Σ by additional T-elements which are enabled in no case of Σ ("dead" T-elements). Thus they do not influence the behaviour of the system. If we associate with each new T-element t a formula $a(t)$, as shown above for events, then $a(t)$ is *valid in* Σ (valid in each case of Σ). In this way it is possible to represent all valid formulae of Σ by a number of "dead" T-elements. Such T-elements are called *facts*.

(d) Definition. Let Σ be a C/E-system.

(i) A formula $a \in A_\Sigma$ is called *valid in* Σ iff for all $c \in C_\Sigma$: $\hat{c}\,(a) = 1$.

(ii) For $B_1, B_2 \subseteq B_\Sigma$, let $t = (B_1, B_2)$ be a new T-element with ${}^{\bullet}t = B_1$ and $t^{\bullet} = B_2$. t is called a *fact* of Σ iff t is never enabled for any $c \in C_\Sigma$.

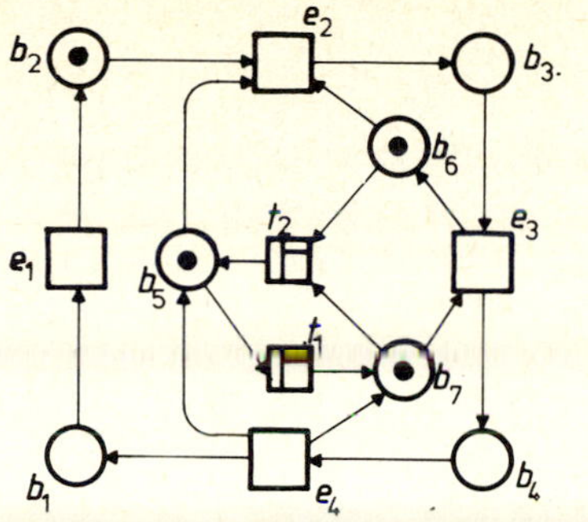

Fig. 52. Enhancement of the system of Fig. 2 by one condition and two facts

In the graphical representation of Σ, a fact $t = (B_1, B_2)$ is drawn as a T-element ⊟ (labelled by a schematic "F"), as already shown in Fig. 51.

For a fact t, the formula $a(t)$ is defined just as $a(e)$ is defined for events e; for instance, if ${}^{\bullet}t = \{b_1, \ldots, b_n\}$, $t^{\bullet} = \{b'_1, \ldots, b'_m\}$, then $a(t) = (b_1 \wedge \ldots \wedge b_n) \rightarrow (b'_1 \vee \ldots \vee b'_m)$.

(e) Theorem. *Let Σ be a finite C/E-system and let $a \in A_{\Sigma}$. a is valid in Σ if and only if facts $t_1, \ldots, t_n$ exist such that a is logically equivalent to $a(t_1) \wedge \ldots \wedge a(t_n)$.*

Proof. Each $a \in A_{\Sigma}$ can be transformed into a logically equivalent formula $a' = a_1 \wedge \ldots \wedge a_k$, where each a_i is a term of the form $\neg b_1 \vee \ldots \vee \neg b_n \vee b'_1 \vee \ldots \vee b'_m$ with $b_i, b'_i \in B_{\Sigma}$ (conjunctive normal form). Therefore, a_i is logically equivalent to a formula $a(t_i)$ with ${}^{\bullet}t_i = \{b_1, \ldots, b_n\}$ and $t_i^{\bullet} = \{b'_1, \ldots, b'_m\}$.

Now, a is valid in $\Sigma \Leftrightarrow a'$ is valid in $\Sigma \Leftrightarrow$ for all i, a_i is valid in $\Sigma \Leftrightarrow$ for all i, $a(t_i)$ is valid in $\Sigma \Leftrightarrow$ for all i, t_i is a fact. □

(f) What about formulae which are valid in some, but not in all, cases of the system? For a case $c \in C_{\Sigma}$, let c' denote the conjunction of all conditions of Σ which hold in c. Then, if a is valid in the cases $c_1, \ldots, c_k$, we can describe this by the valid formula $(c'_1 \wedge \ldots \wedge c'_k) \rightarrow a$.

Exercises for Chapter 4

1. Construct two non-equivalent, contact free C/E-systems Σ and Σ' and a bijection $\varepsilon: E_{\Sigma} \rightarrow E_{\Sigma'}$ such that $\forall e_1, e_2 \in E_{\Sigma}: \sigma(e_1, e_2) = \sigma(\varepsilon(e_1), \varepsilon(e_2))$.

2. Let Σ be a finite, cyclic C/E-system and let $E_1, E_2 \subseteq E_{\Sigma}$. Show that $\sigma(E_1, E_2) = \omega \Leftrightarrow$ there exists a non-empty process $p: K \rightarrow \Sigma$ such that $p({}^{\circ}K) = p(K^{\circ})$ and $v'(p, E_1, E_1) > 0$.

*3. Given a contact-free C/E-system Σ, a process p of Σ, two finite event sets $E_1, E_2 \subseteq E_\Sigma$ and a mapping $g: E_\Sigma \to \mathbb{N}\setminus\{0\}$, let $v_g(p, E_1, E_2) =$

$$\max\left\{\sum_{e \in E_1} g(e) \cdot \mu(p^{-1}(e), D_1, D_2) - \sum_{e \in E_2} g(e) \cdot \mu(p^{-1}(e), D_1, D_2) \,\middle|\, D_1, D_2 \in \underline{\mathrm{sl}}(p)\right\}$$

(*weighted variance* of E_1 and E_2 in p).

Then $\sigma_g(E_1, E_2) = \underline{\sup}\,\{v_g(p, E_1, E_2) \mid p \in \pi_\Sigma\}$ is a *weighted synchronic distance* of E_1 and E_2.

a) Show for all $g: E_\Sigma \to \mathbb{N}\setminus\{0\}$ and all $E_1, E_2, E_3 \subseteq E_\Sigma$:

1) $\sigma_g(E_1, E_2) = 0 \Leftrightarrow E_1 = E_2$,
2) $\sigma_g(E_1, E_2) = \sigma_g(E_2, E_1)$,
3) $\sigma_g(E_1, E_2) \leqq \sigma_g(E_1, E_3) + \sigma_g(E_3, E_2)$.

b) Consider the following C/E-system:

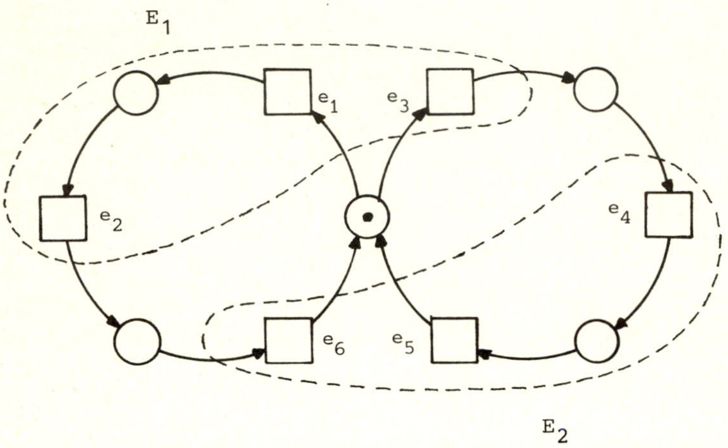

(i) Compute the (unweighted) synchronic distance $\sigma(E_1, E_2)$.
(ii) Does a weight mapping g exist such that $\sigma_g(E_1, E_2)$ is finite?

c) Consider the following C/E-system:

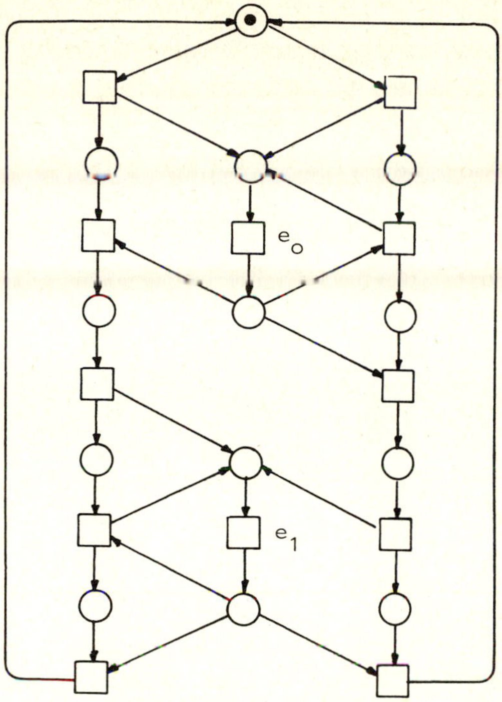

(i) Compute the (unweighted) synchronic distance $\sigma\ (e_1, e_2)$.
(ii) Does a weight mapping g exist such that $\sigma_g\ (e_1, e_2)$ is finite?

4. In the four season system (Fig. 1) represent the following facts:
   a) If it is neither summer nor winter, then it is spring or autumn.
   b) If it is summer then it is neither winter nor autumn.

Part 2. Place/Transition-Nets

As one abstraction of the many ways to interpret nets, we shall consider, in this part, nets with *S*-elements which – in contrast to conditions – may carry more than one token. In such nets *S*-elements are called *places*, the *T*-elements are called *transitions*. An actual state of the system is represented by a certain distribution of tokens over the places, such that the number of tokens on each place is greater than or equal to zero and not greater than its capacity. A transition t may *fire* if all places in $^{\bullet}t$ carry at least one token and if the capacity of all places in $t^{\bullet}$ is greater than the number of tokens they actually carry. When t fires, a token is removed from every place in $^{\bullet}t$ and a token is added to every place in $t^{\bullet}$. We shall also allow *weights* to be attached to the arcs, these weights are natural numbers $n \in \mathbb{N}$. In this case, not one but n tokens are added or removed, respectively, when a transition fires. The firing rule is changed correspondingly; there must be sufficient tokens on each place in $^{\bullet}t$ and sufficient capacity in $t^{\bullet}$ to receive the tokens.

Examples for this kind of nets have already been discussed in Chap. 1 (Fig. 5 and Fig. 6) and also in connection with synchronic distances (Fig. 43).

Chapter 5 explains the basic notions of nets consisting of places and transitions and introduces the coverability graph, a first method for analysing these nets. A further analysis method is the evaluation of invariants which is discussed in Chap. 6. For special classes of nets (free choice nets and marked graphs), analysis methods are derived in Chap. 7.

Chapter 5

Nets Consisting of Places and Transistions

As a first example in this chapter we consider a system consisting of one producer and two consumers. We have already seen this in Fig. 5. In this modified version

(1) the buffer may contain at most five tokens,
(2) the producer generates three tokens in each step,
(3) at most one consumer is able to access the buffer in each configuration of the system,
(4) each consumer removes two tokens when accessing the buffer,
(5) the production steps of the producer are counted.

The system shown in Fig. 53 fulfils these requirements. The meaning of this should be intuitively clear; it is explained formally in the next section.

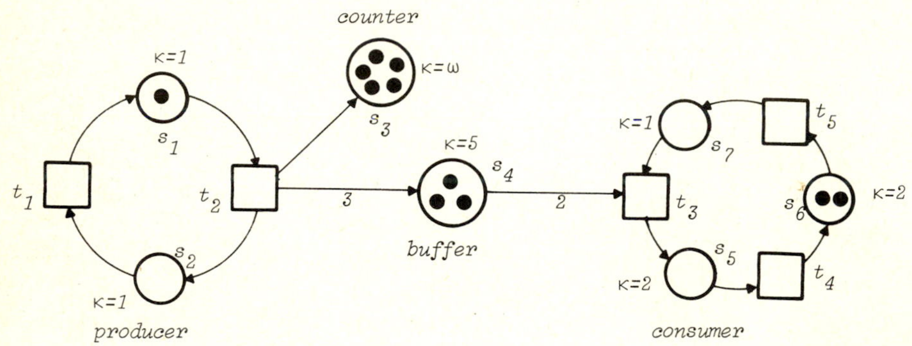

Fig. 53. A producer-consumer system with limited buffer capacity, multiple generation and multiple consumption, limited buffer access, and a counter

5.1 Place/Transition-Nets

This section presents the basic notions of place/transition-nets.

(a) Definition. A 6-tuple $N=(S, T; F, K, M, W)$ is called a *place/transition-net (P/T-net)* iff

(i) $(S, T; F)$ is a finite net, the elements of S and T are called *places* and *transitions*, respectively,

(ii) $K: S \to \mathbb{N} \cup \{\omega\}$, gives a (possibly unlimited) *capacity* for each place,
(iii) $W: F \to \mathbb{N} \setminus \{0\}$, attaches a *weight* to each arc of the net,
(iv) $M: S \to \mathbb{N} \cup \{\omega\}$ is the *initial marking*, respecting the capacities, i.e. $M(s) \leqq K(s)$ for all $s \in S$.

By analogy with *C/E*-systems, the components of a *P/T*-net N are denoted by $S_N, T_N, F_N, K_N, W_N, M_N$.

In the following definition we give the *firing rule* for place/transition-nets.

(b) Definition. Let N be a place/transition-net.

(i) A mapping $M: S_N \to \mathbb{N} \cup \{\omega\}$ is called a *marking of N* iff $M(s) \leq K_N(s)$ for all $s \in S_N$.

Let M be a marking of N.

(ii) A transition $t \in T_N$ is *M-enabled* iff
$\forall s \in {}^{\bullet}t : M(s) \geq W_N(s, t)$ and
$\forall s \in t^{\bullet} : M(s) \leq K_N(s) - W_N(t, s)$.

(iii) An M-enabled transition $t \in T_N$ may yield a *follower marking M' of M* which is such that for each $s \in S_N$

$$M'(s) = \begin{cases} M(s) - W_N(s,t) & \text{iff } s \in {}^{\bullet}t \setminus t^{\bullet}, \\ M(s) + W_N(t,s) & \text{iff } s \in t^{\bullet} \setminus {}^{\bullet}t, \\ M(s) - W_N(s,t) + W_N(t,s) & \text{iff } s \in {}^{\bullet}t \cap t^{\bullet}, \\ M(s) & \text{otherwise.} \end{cases}$$

We say *t fires from M to M'*, and we write $M[t\rangle M'$.

(iv) Let $[M\rangle$ be the smallest set of markings such that
(1) $M \in [M\rangle$ and
(2) if $M_1 \in [M\rangle$ and for some $t \in T_N$ $M_1[t\rangle M_2$ then $M_2 \in [M\rangle$.

In the graphical representation of *P/T*-nets, the arcs $f \in F$ are labelled by $W(f)$ if $W(f) > 1$. The capacity of a place $s \in S$ is represented by the inscription "$\varkappa = K(s)$". The inscription "$\varkappa = \omega$" may be omitted. A marking M is represented by drawing $M(s)$ tokens or the symbol ω on each place s.

Examples of enabled and non-enabled transitions are shown in Fig. 54 and Fig. 55.

Notice that transitions contained in self-loops may only fire if the markings of the corresponding places leave enough latitude (Fig. 56). This is a consequence of the firing rule.

Figure 53 shows a place/transition-net. The marking shown means that the producer must wait for some free place in the buffer, that the consumers com-

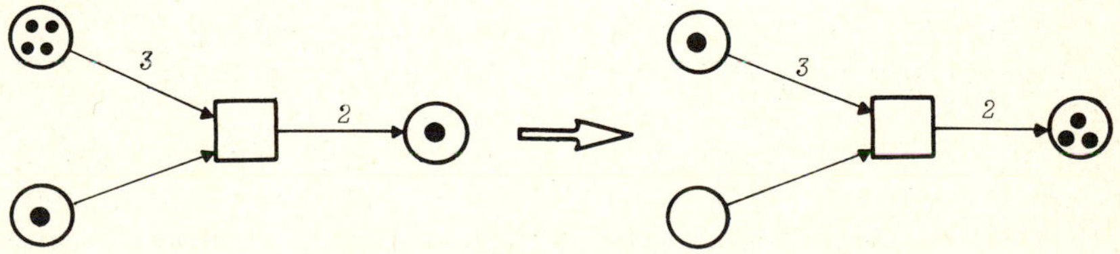

Fig. 54. Firing of a transition

Fig. 55. Situations in which a transition is not enabled

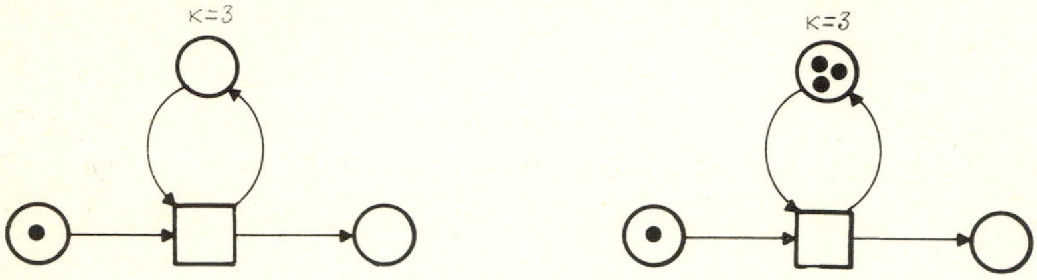

Fig. 56. Both transitions are not enabled and therefore may not fire

pete for the right to access the buffer, and that the producer has already completed five production steps (i.e., it has produced 15 tokens).

Clearly, every C/E-system can be considered as a special place/transition-net with place capacities and arc weights equal to one. Conversely, a place/transition-net with place capacities and arc weights equal to one behaves like a net consisting of conditions and events. But note that a C/E-system is provided with a case class C, whereas for P/T-nets we assume an initial marking.

As a generalization of C/E-systems, a marking M yields a *contact situation* for a transition $t \in T_N$ if t fails to be M-enabled solely because the places in $t^{\bullet}$ do not have sufficient capacity.

(c) Definition. A P/T-net N is called *contact-free* iff for all $M \in [M_N\rangle$ and for all $t \in T_N$:

$$\text{if } \forall s \in {}^{\bullet}t: M(s) \geq W_N(s, t) \text{ then } \forall s \in t^{\bullet}: M(s) \leq K_N(s) - W_N(t, s).$$

Analogously with C/E-systems, every P/T-net can be completed by adding places such that its behaviour is not changed but contact situations are excluded.

Figure 57 shows an example of this construction. Given any P/T-net N, the corresponding net N' is obtained by adding new places and arcs: For every place s of N we construct an additional place $\bar{s}$ and for all arcs (t, s) and (s, t) of F_N we add new arcs $(\bar{s}, t)$ and $(t, \bar{s})$, respectively, such that $W_{N'}(\bar{s}, t) = W_N(t, s)$ and $W_{N'}(t, \bar{s}) = W_N(s, t)$. Assuming the capacity $K_{N'}(\bar{s}) = K_N(s)$ and for the new places $\bar{s}$ the initial marking $M_{N'}(\bar{s}) = K_N(s) - M_N(s)$, the resulting net is obviously contact-free, as for any reachable marking M, $M(s) + M(\bar{s}) = K_N(s)$. Markings M of N and M' of N' *correspond* iff the restriction of M' to the places S_N of N equals M. Obviously, this correspondence is unique. Given

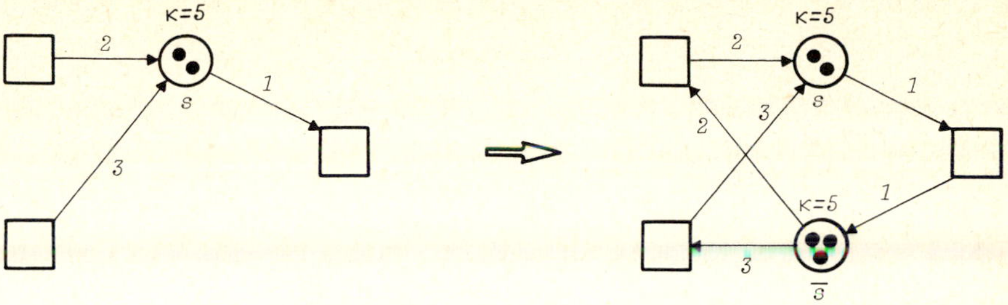

Fig. 57. Complementation in P/T-nets

corresponding marking M of N and M' of N', every transition t is M-enabled in N if and only if t is M'-enabled in N'. Furthermore, we may replace all finite place capacities $K_N(s) \in \mathbb{N}$ in N' by ω without affecting the behaviour of N'.

5.2 Linear Algebraic Representation

The formal treatment of P/T-nets is much simplified by a linear algebraic representation.

(a) Definition. Let $N = (S, T; F, K, M, W)$ be a P/T-net.

(i) For transitions $t \in T$, let the vector $\underline{t}$: $S \to \mathbb{Z}$ be defined as

$$\underline{t}(s) = \begin{cases} W(t,s) & \text{iff } s \in t^{\bullet} \setminus {}^{\bullet}t, \\ -W(s,t) & \text{iff } s \in {}^{\bullet}t \setminus t^{\bullet}, \\ W(t,s) - W(s,t) & \text{iff } s \in {}^{\bullet}t \cap t^{\bullet}, \\ 0 & \text{otherwise.} \end{cases}$$

(ii) Let the matrix $\underline{N}$: $S \times T \to \mathbb{Z}$ be defined as $\underline{N}(s,t) = \underline{t}(s)$.
(Vectors and matrices are introduced in Appendix VII.)

Clearly, every marking of a net may be represented by a vector. Figure 58 shows the matrix $\underline{N}$ and the initial marking M_N of the net shown in Fig. 53. $\underline{N}(s_i, t_j)$ describes the change in the marking of s_i when t_j fires. Entries with value 0 are omitted.

This representation is unambiguous only for pure nets. In this case, the components S_N, T_N, F_N and W_N can be derived. If we additionally require that N is contact-free, the behaviour of N is fully determined by the matrix $\underline{N}$ and the vector M_N.

With this matrix representation we find the following short formulation of the firing rule introduced above:

(b) Corollary. *Let N be a P/T-net and let M, M': $S_N \to \mathbb{N} \cup \{\omega\}$ be two markings of N. Then for each transition $t \in T_N$:*

(i) *If t is M-enabled then $M[t\rangle M' \Leftrightarrow M + \underline{t} = M'$.*

	t_1	t_2	t_3	t_4	t_5	M_N
s_1	1	-1				1
s_2	-1	1				
s_3		1				5
s_4		3	-2			3
s_5			1	-1		
s_6				1	-1	2
s_7			-1		1	

Fig. 58. Matrix and initial marking corresponding to Fig. 53

If N is pure then additionally
(ii) *t is M-enabled* $\Leftrightarrow 0 \leq M + \underline{t} \leq K_N$,
(iii) *N is contact-free* $\Leftrightarrow (\forall M \in [M_N\rangle : 0 \leq M + \underline{t} \Rightarrow M + \underline{t} \leq K_N)$.

For nets with infinite place capacities the following monotonicity property holds:

(c) Lemma. *Let N be a P/T-net with* $\forall s \in S_N$: $K_N(s) = \omega$. *Let* $M_1, M_2 : S_N \to \mathbb{N} \cup \{\omega\}$.
(i) $M_1 [t\rangle M \Rightarrow (M_1 + M_2) [t\rangle (M + M_2)$.
(ii) $M \in [M_1\rangle \Rightarrow (M + M_2) \in [M_1 + M_2\rangle$.

Proof. (i) is obvious from the definitions.
(ii) is implied by (i). □

5.3 Coverability Graphs

It would be nice to have a finite graph directly representing the reachable markings of a (finite) *P/T*-net. Obviously this is impossible, since, in general, infinitely many different markings will be reachable. However, we can get a finite graph such that every reachable marking is either explicitly represented by a node of the graph, or else is "covered" by a node. Therefore such a graph will be denoted *coverability graph.*

In order not to overwhelm the construction we will assume nets N with unlimited capacities, i.e. $K_N(s) = \omega$ for all places $s \in S_N$. According to Sect. 5.1

this is a purely technical restriction, as every P/T-net can be transformed to a net with unlimited capacities without affecting its behaviour.

Each node E of a coverability graph should be thought of as a marking of the net; some will actually be reachable markings, others cover reachable markings. The basic idea of covering markings comes from examining how infinite sequences of reachable markings can arise. One way in which an infinite sequence of distinct markings can arise is as follows. Suppose M and M' are reachable markings and $M' \in [M\rangle$. Suppose further that for each place s $M(s) \leq M'(s)$ and $M \neq M'$ (we write this $M < M'$), and that $K_N(s) = \omega$ at all those places s where $M'(s) > M(s)$; then any transition enabled in M is also enabled in M'. So, by repeating the chain of transitions that lead from M to M' we obtain a new marking M'' with $M' < M''$. Iterating this procedure, we generate an infinite sequence of distinct markings (M_i), $i = 1, 2, \ldots$. Note that this sequence has the property that $M_i(s) = M(s)$ if $M'(s) = M(s)$ while $M_{i+1}(s) > M_i(s)$ if $M'(s) > M(s)$. The sequence will be represented in the graph by a *covering* node K with $K(s) = M(s)$ if $M'(s) = M(s)$ and $K(s) = \omega$ if the number of tokens on s is increasing. Once the construction of the graph is formalised, it will be possible to prove by induction (Lemma (c)) that every reachable marking is either explicitly represented or is covered by such a covering node. Finally, in Theorem (g), we shall prove that only a finite number of nodes are introduced in the construction.

(a) Definition. Let N be a P/T-net with infinite capacities and let $\Gamma = G_0, G_1, \ldots$ be a sequence of graphs which meets the following requirements:

(i) $G_0 = (\{M_N\}, \emptyset)$.

(ii) Let $G_i = (H, P)$ be given. Let $E \in H$ and let $t \in T_N$ such that

(a) t is E-enabled,

(b) no arc starting at E is t-inscribed (i.e. $\nexists E'$ such that $(E, t, E') \in P$).

Then define the marking $\tilde{E}$, for every $s \in S_N$, by $\tilde{E}(s) = \omega$, if there exists a node E' in H such that $E' \leq E + \underline{t}$ and $E'(s) < E(s) + \underline{t}(s)$, and there exists a path from E' to E in G_i, $\tilde{E}(s) = E(s) + \underline{t}(s)$, otherwise, and let $G_{i+1} = (H \cup \{\tilde{E}\}, P \cup \{(E, t, \tilde{E})\})$.

(iii) If it is not possible to construct G_{i+1} following (ii) then let $G_{i+1} = G_i$.

Γ is called a *covering sequence;* $G = \left(\bigcup_{i=0}^{\infty} H_i, \bigcup_{i=0}^{\infty} P_i\right)$ is the *coverability graph generated by* Γ (with $G_i = (H_i, P_i)$).

Notice that, in the above definition, the marking $\tilde{E}$ may already be contained in H, being a node of G_i. In this case only a new arc $(E, t, \tilde{E})$ is added in G_{i+1}, but no new node.

Remember that the assumption of unlimited place capacities is a purely technical restriction. In the following unlimited capacities will be understood if coverability graphs are discussed.

We will now show that indeed each reachable marking is "covered" by a node of a coverability graph:

(b) Lemma. *Let G be a coverability graph of some P/T-net N. For each firing sequence $M_N [t_1\rangle M_1 \ldots M_{n-1} [t_n\rangle M_n$ there exists a path $E_0 t_1 E_1 \ldots E_{n-1} t_n E_n$ in G such that $M_N = E_0$ and for all $i = 1, \ldots, n$, $M_i \leq E_i$.*

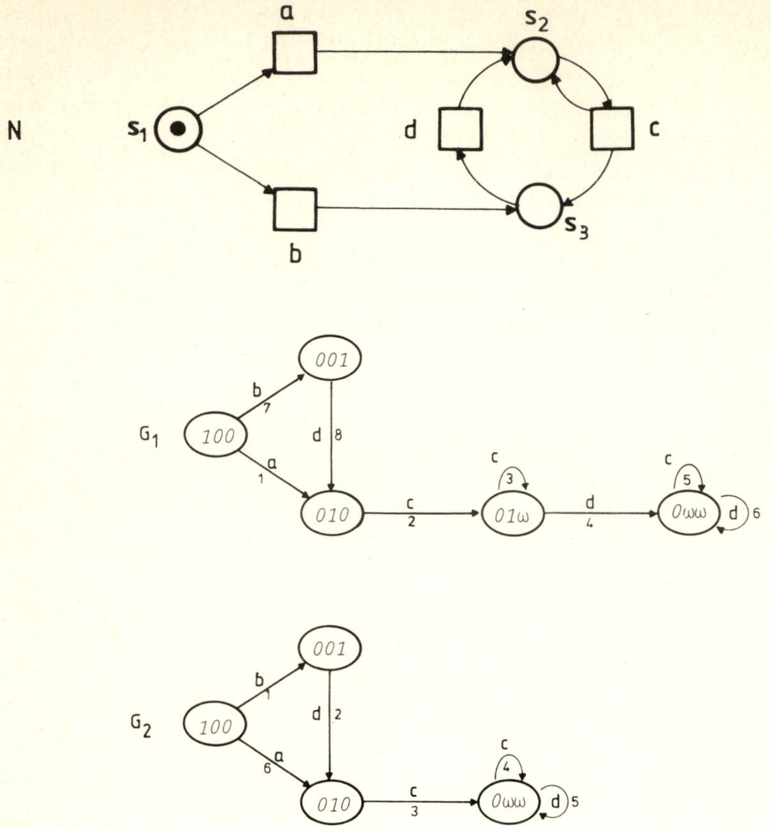

Fig. 59. A P/T-net with two different coverability graphs (Markings M are represented as vectors $M(s_1)\ M(s_2)\ M(s_3)$, arc indices show the order of generation of the arcs)

Proof. We prove the Lemma by induction on n. If $n=0$, $M_0 = M_N$ is by definition a node of G. Assume now there exists a node $E \geq M_{n-1}$. Since t_n is M_{n-1}-enabled, t_n is also E-enabled and there exists an arc (E, t_n, E') in G. Clearly $M_{n-1} + \underline{t_n} \leq E + \underline{t_n} \leq E'$, and the result follows. □

Our next aim is to show that ω-entries in coverability graphs indeed represent unbounded places. This is achieved by associating to each node E of a coverability graph a set of markings such that, for all ω-entries of E, there are infinitely many markings with an unlimited token count on the corresponding place.

(c) Definition. Let N be a P/T-net and let $E: S_N \to \mathbb{N} \cup \{\omega\}$. Let E be a node of G.

(i) Let $\Omega(E) = \{s \in S_N \mid E(s) = \omega\}$.

(ii) For $i \in \mathbb{N}$, a marking M of N is called an *i-marking of* E iff $\forall s \in \Omega(E)$: $M(s) \geq i$ and $\forall s \notin \Omega(E): M(s) = E(s)$.

(iii) Let $\mathcal{M}_E \subseteq [M_N\rangle$ be a minimal set such that, for each $i \in \mathbb{N}$, there exists an i-marking M of E in $\mathcal{M}_E$. Then, $\mathcal{M}_E$ is called a *covering set of* E.

(d) Lemma. *Let G be a coverability graph of some P/T-net, N. For each node E, there exists a covering set* $\mathcal{M}_E$.

Proof. Let $G_0, G_1, \ldots$ be a covering sequence of G. We prove the result by induction following the definition of G.

For the single node of G_0, the proposition is trivially true.

To show the induction step, let $m \in \mathbb{N}$, let $(E, t, \tilde{E})$ be a new arc in G_m, and assume that $\mathcal{M}_E$ exists. We wish to show that $M_{\tilde{E}}$ exists.

Let $E' = E + \underline{t}$. According to the definition of covering sequences, $\Omega(E) \subseteq \Omega(\tilde{E})$. For every set S such that $\Omega(E) \subseteq S \subseteq \Omega(\tilde{E})$ we prove

$$(*) \quad \forall i \in \mathbb{N}\ \exists M \in [M_N\rangle : (\forall s \in S : M(s) \geq i) \wedge (\forall s \notin \Omega(\tilde{E}) : M(s) = E'(s))$$

by induction on $S = \Omega(E), \ldots, S = \Omega(\tilde{E})$.

To show (*) for $S = \Omega(E)$, note that we assume that $\mathcal{M}_E$ exists. As $\Omega(E) = \Omega(E')$, $\mathcal{M}_{E'} = \{M + \underline{t} \mid M \in \mathcal{M}_E\}$ exists. This immediately implies (*) for $S = \Omega(E)$.

By induction hypothesis, assume (*) for some $S = S_1$ and let $s_1 \in \Omega(\tilde{E}) \backslash S_1$. By Definition 5.3 (a) there exists in G_m a node E_0 and a path $E_0\, t_1 \ldots t_n\, E_n$ with $(E_{n-1}, t_n, E_n) = (E, t, \tilde{E})$ such that $E_0 \leq E'$ and $E_0(s_1) < E'(s_1)$. To show (*) for $S_1 \cup \{s_1\}$, let $i \in \mathbb{N}$ and let

$$z = \max(\{|\underline{t_j}(s)| \mid 0 \leq j \leq n \wedge s \in S_1 \wedge \underline{t_j}(s) \leq 0\} \cup \{i\}).$$

By induction hypothesis there exists a marking $M_0 \in [M_N\rangle$ such that $\forall s \in S_1 : M_0(s) \geq (i+1) \cdot n \cdot z$ and $\forall s \notin S_1 : M_0(s) = E'(s)$. Starting with M_0 we can fire from M_0 the transitions $t_1 \ldots t_n$: $M_0[t_1\rangle \ldots [t_n\rangle M_n$, and it holds $\forall s \in S_1 : M_n(s) \geq i \cdot n \cdot z$, $M_n(s_1) > M_0(s_1)$ and $\forall s \notin \Omega(\tilde{E}) : M_n(s) = E'(s)$. So we can even fire $t_1 \ldots t_n$ i times: $M_0[(t_1 \ldots t_n)^i\rangle M$ and for the resulting marking M it holds $\forall s \in S_1 : M(s) \geq n \cdot z$, $M(s_1) \geq i$ and $\forall s \notin \Omega(\tilde{E}) : M(s) = M_0(s)$. This, however, implies (*) for $S = S_1 \cup \{s_1\}$ (as $n \cdot z \geq z \geq i$).

Finally we obtain (*) for $S = \Omega(\tilde{E})$. This expresses the existence of $M_{\tilde{E}}$ and finishes the induction step for G_m. □

These two lemmas motivate the name "coverability graph". Each reachable marking is covered by some node of the graph and, conversely, each node covers a set of reachable markings which may have arbitrarily large values for the ω-components.

Figure 60 shows the kind of structural properties of M_N not represented in the coverability graph. The coverability graph does not show that, in N_1, the transition c may fire arbitrarily often but, in N_2, c may fire at most as many times as a previously fired.

Coverability graphs of finite nets are finite; in Definition 5.3 (a) case (ii) applies only finitely often, as will be shown in the sequel.

(e) Definition. Let N be a P/T-net. Two markings, M_1, M_2, of N are called *unordered* iff neither $M_1 < M_2$ nor $M_2 < M_1$.

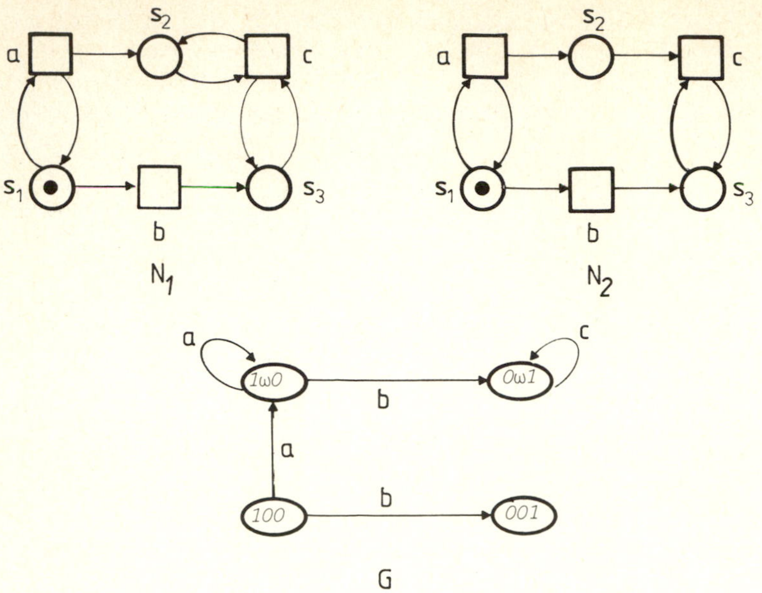

Fig. 60. Two different P/T-nets with the same coverability graph

(f) Lemma. *Every set of pairwise unordered markings of a P/T-net is finite.*

Proof. We prove the somewhat stronger proposition, that each infinite sequence $\sigma = M_1, M_2, \ldots$ of mutually distinct markings has a strongly increasing infinite subsequence $\sigma' = M_{i_1}, M_{i_2}, \ldots$.

If $|S_N| = 1$ then $M_i < M_j$ or $M_j < M_i$ for all $i, j \in \mathbb{N}$. In this case, let $M_{i_1} = M_1$ and, given M_{i_j}, there exist only finitely many markings M in σ such that $M < M_{i_j}$ (as descending sequences of naturals are finite), hence there exists some index $i_{j+1} > i_j$ such that $M_{i_{j+1}} > M_{i_j}$.

For $S_N = \{s_1, \ldots, s_{n+1}\}$, there exists by the induction hypothesis an inifite subsequence $\sigma'' = M_{l_1}, M_{l_2}, \ldots$ of σ such that

(*) $$M_{l_j}(s_k) \leq M_{l_{j+1}}(s_k) \quad \text{for} \quad 1 \leq k \leq n \quad \text{and all} \quad j \in \mathbb{N}.$$

With $M_{i_1} = M_{l_1}$ we construct $\sigma' = M_{i_1}, M_{i_2}, \ldots$ as a subsequence of σ'': Given M_{i_j}, there are only finitely many markings M in σ'' such that $M(s_{n+1}) \leq M_{i_j}(s_{n+1})$. Hence, there exists some index $i_{j+1} > i_j$ such that $M_{i_{j+1}}$ in σ'' and $M_{i_{j+1}}(s_{n+1}) > M_{i_j}(s_n)$. With (*), we have $M_{i_{j+1}} > M_{i_j}$. □

(g) Theorem. *Every coverability graph of a P/T-net is finite.*

Proof. For $j = 1, 2, \ldots$, let (K_{j-1}, t_j, K_j) be the arc which was added in G_j. Let $\Gamma_0 = G_0, G_1, \ldots$ be a covering sequence of a finite P/T-net and let G be the coverability graph generated by Γ. A path $w = K_0\, t_1\, K_1 \ldots$ of G is called *constructive* iff there exists a subsequence $G_{i_0}, G_{i_1}, \ldots$ of Γ such that G_{i_j} generates the arc (K_{j-1}, t_j, K_j) $(j=1, 2, \ldots)$ and $G_{i_0} = G_0$. We shall show that every constructive path $w = K_0\, t_1\, K_1 \ldots$ is finite. Let $\Phi = K_0, K_1, \ldots$ be the sequence of nodes in w and let $S = \{s \in S_N \mid M_N(s) \neq \omega\}$. For each descending subsequences $K_0 > K_1' > \ldots > K_n'$, $n \leq \sum_{s \in S} K_0(s)$. For each increasing subsequence $K_0' < K_1' < \ldots < K_n'$, we have by construction of w, $K_i'(s) < K_j'(s) \Rightarrow K_j'(s) = \omega$.

Therefore $n \leq |S_N|$. Hence Φ and also w is finite.

Obviously the constructive paths of G constitute an acyclic subgraph G' of G. As G' is finitely based and finitely branched, and as each constructive path is finite, G' is finite according to Koenig's Lemma (cf. A16). Since every node of G lies on some constructive path, the node sets of G and of G' are equal and the Theorem follows. □

Thus, coverability graphs can actually be constructed for P/T-nets and can be used to prove certain properties of such nets.

5.4 Decision Procedures for Some Net Properties

Some questions about coverability and liveness can be reduced to properties of coverability graphs. Since coverability graphs of P/T-nets are finite and can actually be constructed, we obtain constructive procedures for the decision of these problems. Such procedures are the main concern of this section.

It is decidable for arbitrary markings M of a P/T-net N whether a marking $M' \in [M_N\rangle$ with $M \leq M'$ exists, that is, M is *covered* by some marking of $[M_N\rangle$:

(a) Theorem. *Let N be a P/T-net, let $M: S_N \to \mathbb{N} \cup \{\omega\}$ be an arbitrary marking of N and let G be a coverability graph of N. A marking $M' \in [M_N\rangle$ with $M \leq M'$ exists if and only if*

(i) $\forall s \in S_N : (M(s) = \omega \Rightarrow M_N(s) = \omega)$ *and*

(ii) *there exists a node E in G such that $M \leq E$.*

Proof. Let $M' \in [M_N\rangle$ with $M \leq M'$. (i) using Lemma 5.3 (b), there exists a node E of G with $M' \leq E$. Therefore, $M \leq E$. (ii) Clearly, $M_N(s) \neq \omega$ implies $\forall M' \in [M_N\rangle : M'(s) \neq \omega$.

Conversely, assume (i) and (ii), let E be a node of G with $M \leq E$. Using Lemma 5.3 (d), there exists $M' \in [M_N\rangle$ with $M'(s) \geq M(s)$ in the case $E(s) \in \mathbb{N}$, and M'(s) arbitrarily large in the case $E(s) = \omega$. If $M(s) = \omega$, we have $M_N(s) = \omega$ and therefore $M'(s) = \omega$. □

(b) Definition. Let N be a P/T-net. $S \subseteq S_N$ is called *simultaneously unbounded* iff $\forall i \in \mathbb{N}\ \exists M_i \in [M_N\rangle$ such that $\forall s \in S : M_i(s) \geq i$.

(c) Theorem. *Let N be a P/T-net, let $S \subseteq S_N$ and let G be a coverability graph of N. S is simultaneously unbounded iff there exists a node E in G such that $\forall s \in S : E(s) = \omega$.*

Proof. Let $M_1, M_2, \ldots \in [M_N\rangle$ such that $\forall s \in S\ \forall i \in \mathbb{N} : M_i(s) \geq i$. Using 5.3 (b) there exists, for each M_i, a node E_i such that $M_i \leq E_i$. Since G is finite (5.3 (g)), there exists a node E of G such that, for infinitely many $i_1, i_2, \ldots \in \mathbb{N}$, $M_{i_j} \leq E$. Since $\forall s \in S : i_j \leq M_{i_j}(s) \leq E(s)$, we have $E(s) = \omega$.

The converse is Lemma 5.3 (d). □

(d) Definition. Let N be a P/T-net, let M: $S_N \to \mathbb{N} \cup \{\omega\}$ be a marking of N, and let $t \in T_N$.

t is called *M-dead* iff $\forall M' \in [M\rangle : t$ is not M'-enabled.

(e) Theorem. *Let N be a P/T-net, let $t \in T_N$ and let G be a coverability graph of N.*

t is M_N-dead iff there exists no arc of the form (E, t, E') in G.

Proof. If $(E, t, \tilde{E})$ is an arc of G then $E[t\rangle\ \tilde{E}$ and, by Lemma 5.3 (d) there exists a marking $M \in \mathscr{M}_E$ which enables t.

If t is not M_N-dead then there exist $M_1, M_2 \in [M_N\rangle$ with $M_1[t\rangle M_2$, So, by Lemma 5.3 (b) there exists a node E with $M_1 \leq E$. Since t is M_1-enabled, it is also E-enabled and an arc $(E, t, \tilde{E})$ exists. □

(f) Theorem. *Let N be a P/T-net, such that $\forall s \in S_N$: $K_N(s) = \omega$, let M: $S_N \to \mathbb{N} \cup \{\omega\}$ be a marking of N, and let $t \in T_N$ be M-dead. Then for all $M' < M$, t is M'-dead.*

Proof. Assume t is not M'-dead. Then there exists a marking $\tilde{M}' \in [M'\rangle$ such that t is $\tilde{M}'$-enabled. Starting from M, firing the same transitions in the same order as when firing from M' to $\tilde{M}'$, yields a marking $\tilde{M}$ such that t is $\tilde{M}$-enabled. □

(g) Corollary. *Let N be a P/T-net and let G be a coverability graph of N. The set $[M_N\rangle$ of reachable markings is finite iff no node of G has an ω-component.*

Proof. $[M_N\rangle$ is infinite iff at least one place s is unbounded. According to Theorem 5.4 (c), this is true iff at least for one node E of G, $E(s) = \omega$. □

For the practical analysis of nets, coverability graphs are of limited value, as algorithms for their construction are too complex. It was shown in [81] (cf. also [47, 73]) that there exists a sequence $N_1, N_2, \ldots$ of P/T-nets with linearly growing size (let the size of a net be the number of its elements, arcs, and initial tokens) such that the corresponding coverability graphs $G_0, G_1, \ldots$ grow (with respect to the number of nodes) quicker than any primitive recursive function.

As a consequence of this result, the following is proved in [81] and [82]: Let N and N' be two P/T-nets with identical places (i.e. $S_N = S_{N'}$) and finite sets $[M_N\rangle$ and $[M_{N'}\rangle$ of reachable markings. It is obviously decidable if $[M_N\rangle \subseteq [M_{N'}\rangle$, but not in primitive recursive time (or space). A similar result holds for the problem whether or not $[M_N\rangle = [M_{N'}\rangle$.

Assuming N and N' as above, but with infinite sets $[M_N\rangle$ and $[M_{N'}\rangle$, the problems $[M_N\rangle \subseteq [M_{N'}\rangle$ and $[M_N\rangle = [M_{N'}\rangle$ are not decidable [76]. Furthermore it is shown there that it is not decidable if $[M_N\rangle$ decreases in case a transition is skipped from the net.

For a P/T-net N it is decidable in space $2^{c \cdot n \cdot \log(n)}$ (let n denote the size of N) if $[M_N\rangle$ is finite [80]. Hence the construction of coverability graphs is not required for this problem. Equally complex is the problem if, for an arbitrary marking M, there exists a reachable marking $M' \in [M_N\rangle$ such that $M \leq M'$. Furthermore it is shown in [80] that both problems can not be decided in space $2^{\sqrt{n}}$.

The problem if an arbitrary marking M of a P/T-net N is a reachable marking (i.e. $M \in [M_N\rangle$) became well known as the *reachability problem*. It was recently (positively) solved [67].

5.5 Liveness

P/T-nets are often used in application areas where the number and distribution of dynamically moving objects is important; for instance, the data in a computer, the goods in a warehouse, the documents in an administration system, the work in progress in a production system. In such areas, the aim is generally to obtain an organisation which allows for variations in the number and distribution of the moving objects, but which restricts such variations within certain limits. There may be failures in the form of blockings, which cause a partial or total standstill of the system. Such blockings are either the result of a lack of such moving objects, or the result of a jam (superfluity).

In the net representation of such systems, active system elements (processors, agents, machines) are represented as transitions, passive system elements (buffers, stores) are represented as places. Moving objects are represented as tokens. Then, blockings are visible as transitions which are not able to fire any more. Such nets are not *live*. There are several notions of liveness; a marking may be called live if, for each follower marking, there exists some enabled transition, or if each transition may sometimes be enabled, or if each transition may sometimes be enabled from each follower marking, or if each (or at least one) follower marking is reproducible, etc. A net may be called live if, with respect to any of the above liveness notions for markings, it can be provided with a live marking.

In the following we use a notion of liveness which requires, for each marking, the possibility of each transition being enabled.

(a) Definition. Let N be a P/T-net, let $t \in T_N$.
(i) t is called *live* iff $\forall M \in [M_N\rangle\ \exists M' \in [M\rangle$ such that t is M'-enabled.
(ii) N is called *live* iff $\forall t \in T_N$: t is live.

The intuitively obvious conjecture that enlarging (adding tokens to) the initial marking of a live net yields again a live net turns out to be false. Figure 61 shows a counterexample.

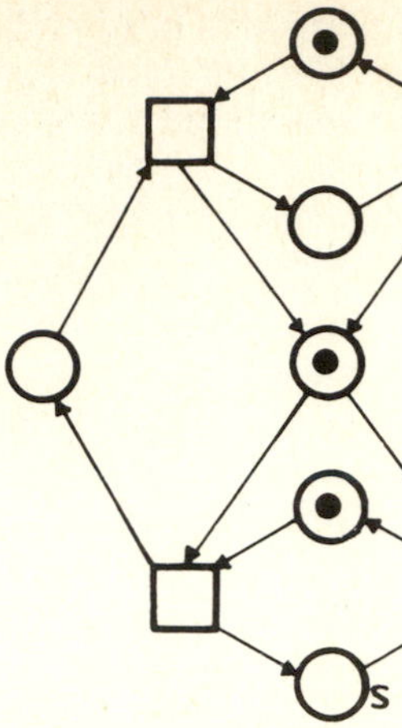

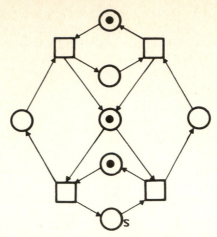

Fig. 61. A live P/T-net. If, additionally, the place s is marked, this yields a net which is no longer live

This liveness notion does not imply that each marking is reproducible, i.e. for all $M_1, M_2 \in [M_N\rangle$: $M_2 \in [M_1\rangle$. Even then this is not the case, if all capacities are finite. An example of this is shown in Fig. 24.

It might be interesting to discuss liveness of markings:

(b) Definition. A marking M of a P/T-net N is *live* iff $\forall t \in T_N\ \exists M' \in [M\rangle$ such that t is M'-enabled.

Then we get the following.

(c) Lemma. *A P/T-net N is live iff all markings $M \in [M_N\rangle$ are live.*

Proof. N is live $\Leftrightarrow \forall t \in T_N : t$ is live $\Leftrightarrow \forall t \in T_N\ \forall M \in [M_N\rangle\ \exists M'$ such that t is M'-enabled $\Leftrightarrow \forall M \in [M_N\rangle\ M$ is live. □

Exercises for Chapter 5

1. Consider the P/T-net of Fig. 12.
   a) Introduce minimal capacities which do not affect the behaviour of the net.
   b) Construct the matrix of this net.

2. Construct different coverability graphs for the following P/T-net:

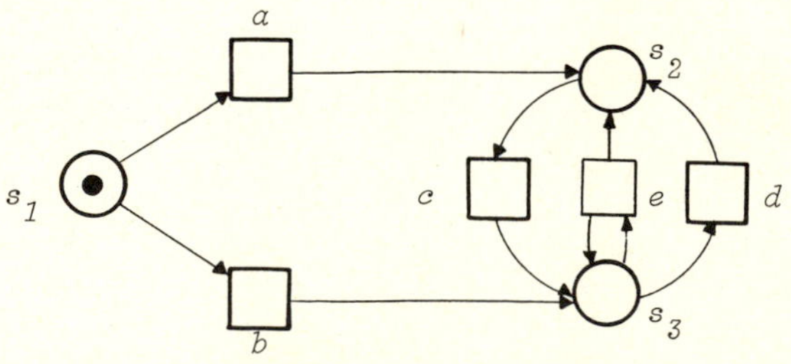

3. Construct a P/T-net with three different coverability graphs.

4. Construct three different P/T-nets with equal coverability graphs.

5. Show that in the P/T-net of Exercise 2
   a) $\exists M' \in [M_N\rangle$ with $(0, 5, 10) < M'$,
   b) $\nexists M' \in [M_N\rangle$ with $(1, 2, 3) < M'$,
   c) $\{s_2, s_3\}$ is simultaneously unbounded,
   d) there exist no M_N-dead transitions.

6. Consider the following P/T-net:

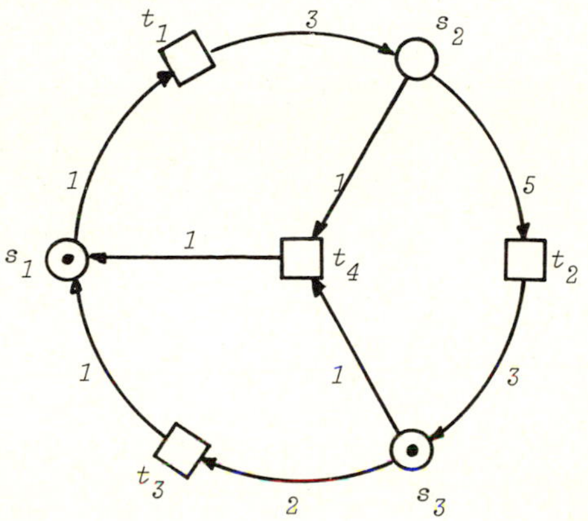

   a) Which subsets of places are simultaneously unbounded?
   b) Is the net live?
   (Hint: Construct the coverability graph.)

7. Is the net in Exercise 2 live?

8. Rearrange Fig. 12 such that never only one process is reading. If two processes are reading, a third one may join them.

*9. a) In Sect. 4.1 (h) we suggested a graphical representation for synchronic distances. Formalise this idea.
   Hints. Let a C/E-system Σ and a pair $s = (E_1, E_2)$ of subsets of E_Σ be given.
   Define the net Σ_s by $\Sigma_s = (B_\Sigma \cup \{s\}, E_\Sigma; F_\Sigma \cup \{(e,s) \mid e \in E_1\} \cup \{(s,e) \mid e \in E_2\})$.
   Together with an initial marking M, Σ_s can be conceived as a P/T-net.
   Define now what it means to simulate an (initial part of a) process p: $K \to \Sigma$ as a firing sequence in Σ_s. To do this, consider event sequences which are obtained by extending the partial order of T_K to total orders: Let $w = e_1 \ldots e_n \in \mathscr{S}(p)$ iff there exists a slice D_w of K *with* $D_w^- \cap T_K = \{t_1, \ldots, t_n\}$ such that for all $1 \le i, j \le n$, $p(t_i) = e_i$ and $(t_i < t_j \Rightarrow i < j)$.
   As an example, for the process p as shown in Fig. 41, we obtain $\mathscr{S}(p) = \{e_0\, e_1\, e_2\, e_3,\ e_0\, e_2\, e_1\, e_3,\ e_0\, e_1\, e_2,\ e_0\, e_2\, e_1,\ e_0\, e_2,\ e_0\}$.

Let $w = e_1 \dots e_n \in \mathcal{S}(p)$ and let D_w be a slice as given in the definition of $\mathcal{S}(p)$. For $E \subseteq E_\Sigma$ let $\lambda(E, w) = \{i \mid e_i \in E\}$. Obviously, $\lambda(E, w) = |p^{-1}(E) \cap D_w^-|$.
If w is embedded in a firing sequence $M_0 [e_1\rangle M_1 \dots M_{n-1} [e_n\rangle M_n$ of Σ_s, $\tilde{\mu}(w, s) = \lambda(E_1, w) - \lambda(E_2, w)$ denotes the effect of w to s, i.e. $\tilde{\mu}(w, s) = M_n(s) - M_0(s)$ (as obviously $M_n(s) = M_0(s) + \lambda(E_1, w) - \lambda(E_2, w)$).
$\tilde{\mu}(w, s)$ is the contribution of w to the variance $\tilde{v}$ of p, defined by $\tilde{v}(p, s) = \max\{\tilde{\mu}(w, s) \mid w \in \mathcal{S}(p)\} - \min\{\tilde{\mu}(w, s) \mid w \in \mathcal{S}(p)\}$. $\tilde{v}(p, s)$ defines the contribution of p to the maximal variation of the number of tokens on s.
Define now $\tilde{\sigma}(E_1, E_2) = \sup\{\tilde{v}(p, s) \mid p \in \pi_\Sigma\}$ and show $\tilde{\sigma} = \sigma$. (Obviously it is sufficient to show $\tilde{v}(p, s) = v(p, E_1, E_2)$).

b) Let Σ, Σ_s, and $\mathcal{S}(p)$ be as above and let the set $\mathfrak{M}$ of markings of Σ_s be defined by: $M \in \mathfrak{M}$ iff $M(s) \in \mathbb{N}$ and there exists a case $c \in C_\Sigma$ such that $\forall b \in B_\Sigma : M(b) = 1$ if $b \in c$ and $M(b) = 0$ if $b \notin c$.
Let $\tilde{\tilde{\sigma}}(E_1, E_2) = \sup\{M_n(s) - M'_{n'}(s) \mid \exists p \in \pi_\Sigma \; \exists M \in \mathfrak{M}$ such that there exist two firing sequences $M[a_1\rangle M_1 \dots M_{n-1}[a_n\rangle M_n$ and $M[a'_1\rangle M'_1 \dots M'_{n'-1}[a'_{n'}\rangle M'_{n'}$ with $\{a_1 \dots a_n, a'_1 \dots a'_{n'}\} \subseteq \mathcal{S}(p)\}$. Show that σ and $\tilde{\tilde{\sigma}}$ are equal.

Chapter 6

Net Invariants

In this chapter, we are first concerned with sets of places of P/T-nets which do not change their token count during transition firings. Knowledge about any such sets of places not only helps in analysing liveness but also allows us to investigate other properties of systems (for instance, facts in C/E-systems). Such sets of places will be called *S-invariants*. Since invariants are characterized by solutions of linear equation systems of the form $\underline{N}' \cdot x = 0$, ($\underline{N}'$ denotes the transpose of N, cf. Appendix VII) it is possible to compute them by the well-known methods of linear algebra.

By means of two examples, a sender-receiver model and a seat-reservation system, we shall discuss how to apply invariants to the construction and analysis of systems.

As well as S-invariants, we also obtain *T-invariants* as solutions of $\underline{N} \cdot x = 0$. They indicate how often, starting from some marking, each transition has to fire, to reproduce this marking.

6.1 *S*-Invariants

To begin with we shall consider a special class of S-invariants. Let N be a P/T-net with arcweight 1 for all arcs. We want to characterize sets of places, $S \subseteq S_N$, of N which do not change their joint total token count when transitions fire. Certainly we can see that if S is such a set of places and $s \in S$ then for each transition $t \in s^{\bullet}$ which may be enabled there must be a place $s' \in t^{\bullet}$ which is also contained in S. Intuitively speaking, a token flows along the arcs (s, t) and (t,s') from s to s'. Analogously, there is, for each transition $t \in {}^{\bullet}s$ which may be enabled, a place $s' \in {}^{\bullet}t$ such that a token flows along (s', t) and (t, s) from s' to s. Thus, S may be characterized by a set, F, of arcs which fulfils the following requirements:

1) When an arc belonging to F starts or ends at a place s then all arcs from and to s belong to F.
2) For each arc of F ending at some transition t there is exactly one arc belonging to F starting at t.

Figure 62 shows such a set of places. The corresponding arcs are represented by thick lines. The token count is also constant on the set of places $\{s_1, s_2, s_4, s_5\}$.

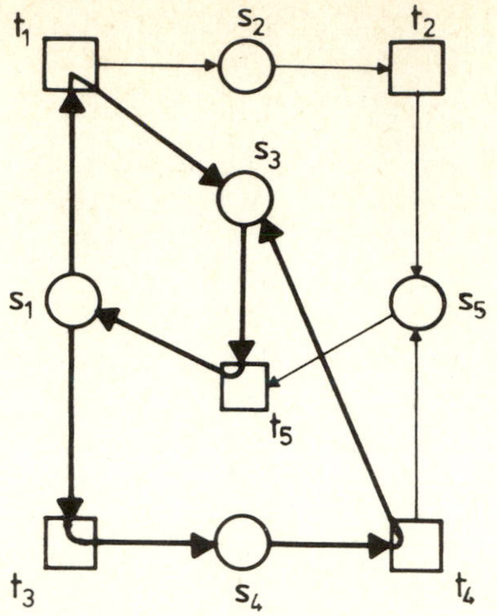

Fig. 62. The sum of tokens on the set $\{s_1, s_3, s_4\}$ of places is not changed by transition firings

This simple method of characterizing sets of places with constant token count does not work if there are arcweights other than 1. An example is shown in Fig. 63. Therefore, we have to investigate further how the firing of transitions affects such sets of places.

If the token count on $S \subseteq S_N$ does not change when a transition $t \in T_N$ fires then

$$\sum_{s \in {}^{\bullet}t \cap S} W(s, t) = \sum_{s \in t^{\bullet} \cap S} W(t, s).$$

By Definition 5.2 (a), this condition is equivalent to

$$\sum_{s \in {}^{\bullet}t \cap S} \underline{t}(s) = -\sum_{s \in t^{\bullet} \cap S} \underline{t}(s), \text{ i.e. } \sum_{s \in {}^{\bullet}t \cap S} \underline{t}(s) + \sum_{s \in t^{\bullet} \cap S} \underline{t}(s) = 0.$$

This is equivalent to

$$\sum_{s \in ({}^{\bullet}t \cup t^{\bullet}) \cap S} \underline{t}(s) = 0 \quad \text{and even to} \quad \sum_{s \in S} \underline{t}(s) = 0.$$

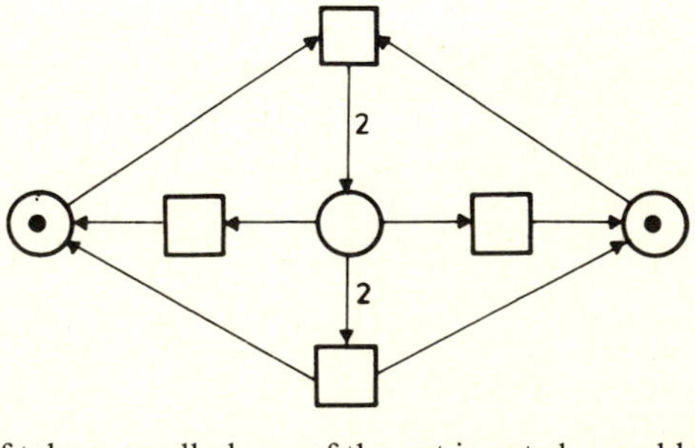

Fig. 63. The sum of tokens on all places of the net is not changed by transition firings

If we replace S by its characteristic vector c_S (see A20) the condition becomes

$$\sum_{s \in S_N} \underline{t}(s) \cdot c_S(s) = 0 \text{ or, by vector multiplication, } \underline{t} \cdot c_S = 0.$$

If the token count on $S \subseteq S_N$ never changes under arbitrary transition firings, the condition $\underline{t}_i \cdot c_S = 0$ must be fulfilled for all transitions $t_i \in T_N$, hence $\underline{N}' \cdot c_S = 0$ must hold.

Conversely, each solution c of $\underline{N}' \cdot x = 0$ consisting of components from $\{0, 1\}$ is a characteristic vector of a set of places with constant token count. So such sets are found by solving $\underline{N}' \cdot x = 0$.

We shall now make precise this informally introduced relation between sets of places with constant token count and solutions of linear equations, and introduce the general class of S-invariants.

(a) Definition. Let N be a P/T-net.
A place vector i: $S_N \to \mathbb{Z}$ is called an *S-invariant of N* iff $\underline{N}' \cdot i = 0$.

(b) Lemma. *Let i_1 and i_2 be S-invariants of a net N and let $z \in \mathbb{Z}$. Then $i_1 + i_2$ and $z \cdot i_1$ are also S-invariants of N.*

Figure 64 shows invariants of the net of Fig. 62. The only invariants which are characteristic vectors are i_1 and i_2. In fact, they denote the sets $\{s_1, s_3, s_4\}$ and $\{s_1, s_2, s_4, s_5\}$, which we previously recognized as sets of places with a constant token count.

How can we now interpret the S-invariants which are not characteristic vectors? The token count on the corresponding places is certainly not constant, but on the other hand it does not vary without limit. Considering Fig. 62, we can say that a token on s_1 "counts" as much as a token on s_2 and a token on s_3 together. Similarly, a token on s_4 "counts" as much as two tokens distributed on s_3 and s_5. Tokens on s_1 and s_4 have a "weight", which is twice that of tokens

	t_1	t_2	t_3	t_4	t_5	i_1	i_2	i_3	i_4
s_1	-1		-1		1	1	1	2	
s_2	1	-1					1	1	1
s_3	1			1	-1	1		1	-1
s_4			1	-1		1	1	2	
s_5		1		1	-1		1	1	1

Fig. 64. The matrix and four invariants of the net shown in Fig. 62

on s_2, s_3 and s_5. If we consider these weights we find "weighted" token counts on the net which remain constant during transition firings: Let M_1 and M_2 be markings of the net of Fig. 62 and let $t \in \{t_1, \ldots, t_5\}$ be a transition such that $M_1\,[t\rangle\, M_2$.

Then,

$$2M_1(s_1) + 2M_1(s_4) + M_1(s_2) + M_1(s_3) + M_1(s_5) =$$
$$2M_2(s_1) + 2M_2(s_4) + M_2(s_2) + M_2(s_3) + M_2(s_5).$$

So, with invariant i_3 of Fig. 64:

$$M_i \cdot i_3 = M_2 \cdot i_3.$$

Considering again Fig. 62, we find a further regularity concering the places s_2, s_3 and s_5. s_2 and s_3 always get (by t_1) the same number of tokens. The tokens of s_2 may flow to s_5. From s_5 and s_3 the same number of tokens is always removed (by t_5). Hence the token count on s_3 varies in the same way as the sum of tokens on s_2 and s_5. Therefore, $M(s_3) = M(s_2) + M(s_5)$ for all reachable markings $M \in [M_0\rangle$ of a marking M_0 with $M_0(s_2) = M_0(s_3) = M_0(s_5) = 0$. Using invariant i_4 of Fig. 64 we have $M_0 \cdot i_4 = 0 = M \cdot i_4$.

(c) Lemma. *Let N be a P/T-net with a positive S-invariant i and let* $S = \{s \in S_N \mid i(s) > 0\}$.
Then $S^{\bullet} = {}^{\bullet}S$.

Proof. Assume there exists $t \in S^{\bullet} \setminus {}^{\bullet}S$. Then

$$\exists s \in S : \underline{t}(s) < 0 \quad \text{and} \quad \forall s \in S : \neg(\underline{t}(s) > 0).$$

Then clearly $\underline{t} \cdot c_S < 0$ and, since i is positive, $c_S \leq i$ and therefore $\underline{t} \cdot i < 0$. So i is, under this assumption, not an S-invariant. For $t \in {}^{\bullet}S \setminus S^{\bullet}$, we find similarly $\underline{t} \cdot i > 0$. □

This corollary corresponds to the intuition that sets of places with constant token count are obtained from sets of arcs which lead from a place in ${}^{\bullet}t$ to a place in $t^{\bullet}$.

(d) Theorem. *Let N be a P/T-net. Then, for each S-invariant i of N and each reachable marking* $M \in [M_N\rangle$, $M \cdot i = M_N \cdot i$.

Proof. Let $M_1, M_2 \in [M_N\rangle$ and let $t \in T_N$ such that $M_1\,[t\rangle\, M_2$. Then, in particular, $M_2 = M_1 + \underline{t}$ (Corollary 5.2 (b)) and $\underline{t} \cdot i = 0$ (since i is an invariant). Therefore $M_2 \cdot i = (M_1 + \underline{t}) \cdot i = M_1 \cdot i + \underline{t} \cdot i = M_1 \cdot i$. □

The converse of this theorem is only true if every transition may fire at least once; in particular, it is true for live nets.

(e) Lemma. *Let N be a live P/T-net and let* $i : S_N \to \mathbb{Z}$ *be a place vector such that, for all* $M \in [M_N\rangle$, $M \cdot i = M_N \cdot i$. *Then i is an S-invariant.*

Proof. It is sufficient to show, for each transition $t \in T_N$, $\underline{t} \cdot i = 0$. So let $t \in T_N$ and let $M \in [M_N\rangle$ such that t is M-enabled. Then, with $M[t\rangle M'$, $M \cdot i = M' \cdot i = (M + \underline{t}) \cdot i$ (Corollary 5.2 (b)) $= M \cdot i + \underline{t} \cdot i$. Hence $\underline{t} \cdot i = 0$. □

(f) Corollary. *Let N be a live P/T-net and let $i : S_N \to \mathbb{Z}$ be a place vector. i is an S-invariant if and only if for all $M \in [M_N\rangle$, $M \cdot i = M_N \cdot i$.*

(g) Corollary. *Let N be a P/T-net and let $S \subseteq S_N$ be a set of places whose characteristic vector c_S is an S-invariant.*
Then, for all $M \in [M_N\rangle$, $\sum_{s \in S} M(s) = \sum_{s \in S} M_N(s)$.

6.2 Nets Covered by S-Invariants

If a place s of a P/T-net N may obtain unboundedly many tokens then s may not belong to any positive invariant i. This section deals with this dependency between the boundedness of places and their being contained in invariants.

(a) Definition. A P/T-net N is said to be *covered by S-invariants* iff, for each place $s \in S_N$, there exists a positive S-invariant i of N with $i(s) > 0$.

(b) Corollary. *If some P/T-net N is covered by S-invariants then there exists an invariant i with $i(s) > 0$ for all $s \in S_N$.*

Proof. By the hypothesis, there exists, for each $s \in S_N$, an invariant i_s with $i_s(s) > 0$. Using Corollary 6.1 (b), $i = \sum_{s \in S_N} i_s$ is an invariant fulfilling the requirements. □

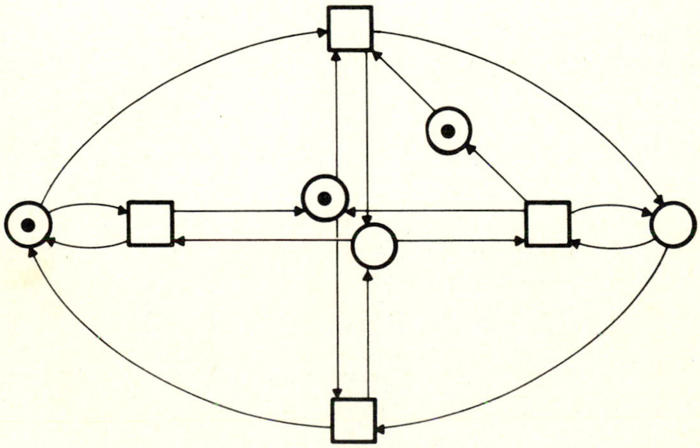

Fig. 65. This net, which is live and contact-free with capacity 1, is not covered by S-invariants

(c) Definition. A *P/T*-net N is called *bounded* iff M_N is finite and there exists $n \in \mathbb{N}$ such that, for all $M \in [M_N\rangle$ and all $s \in S_N$, $M(s) \leq n$.

(d) Theorem. *Let N be a P/T-net and let M_N be finite. If N is covered by S-invariants then N is bounded.*

Proof. Let $s_0 \in S_N$ and let i be a positive S-invariant with $i(s_0) > 0$; let $M \in [M_N\rangle$. Since $M(s_0) \cdot i(s_0) \leq \sum_{s \in S_N} M(s) \cdot i(s) = M \cdot i = M_N \cdot i$. (Theorem 6.1 (d)), we have $M(s_0) \leq \dfrac{M_N \cdot i}{i(s_0)}$. □

The converse of this theorem is not true, even if N is presupposed to be live or if the limit for the number of tokens is assumed to be one. Figure 65 shows such a net.

6.3 The Verification of System Properties Using *S*-Invariants

We first consider a small example to show which structural properties can be recognized by a knowledge of the *S*-invariants of a net. Suppose that n processes in an operating system are each allowed to access a buffer in reading or writing mode. To guarantee reliability, reading and writing access is restricted in the following way: When no process is writing to the buffer then up to $k \leq n$ processes are allowed to read it. But writing access to the buffer is only permitted as long as no other process is reading or writing the buffer.

In Fig. 66, such a system of reader and writer processes is shown as a *P/T*-net. Each process is in one of five states, represented by the places $s_0, \ldots, s_4$. In the initial state, all n processes are passive; hence s_0 contains n tokens under the initial marking M_N. The place s_5 contains k tokens in M_N. This

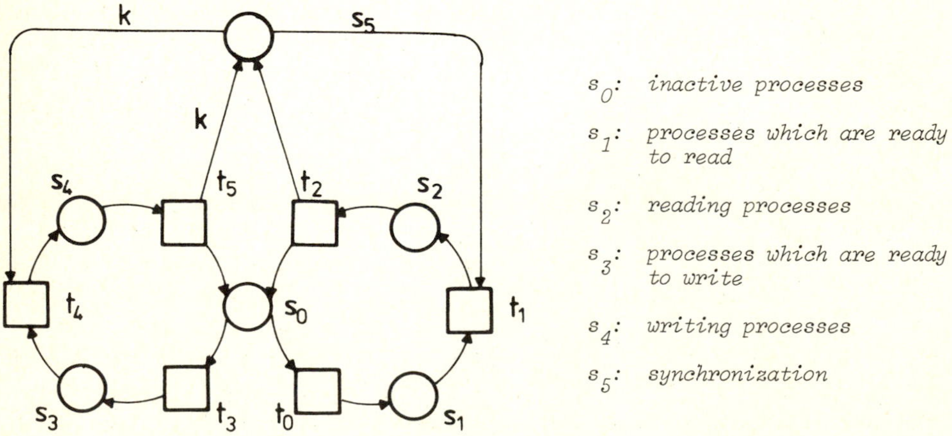

Fig. 66. A system of reader and writer processes of an operating system

	t_0	t_1	t_2	t_3	t_4	t_5	i_1	i_2	M_N
s_0	-1		1	-1		1	1		n
s_1	1	-1					1		
s_2		1	-1				1	1	
s_3				1	-1		1		
s_4					1	-1	1	k	
s_5		-1	1		$-k$	k		1	k

Fig. 67. Matrix, invariants i_1, i_2 and initial marking of the net shown in Fig. 66

corresponds to the number of processes which are allowed to read the buffer concurrently.

With the invariants shown in Fig. 67, it is possible to prove the correctness of the system design.

Using i_1, we have, for each follower marking $M \in [M_N\rangle$:

$$\sum_{i=0}^{4} M(s_i) = \sum_{i=0}^{4} M_N(s_i) = n.$$

This means: The number, n, of processes remains constant and each process is in one of the states $s_0, \ldots, s_4$.

Using i_2, we have, for each marking $M \in [M_N\rangle$:

$$M(s_2) + k \cdot M(s_4) + M(s_5) = M_N(s_2) + k \cdot M_N(s_4) + M_N(s_5) = k.$$

Hence, we find: s_4 contains at most one token under M; that is, there exists at most one writing process. When s_4 carries a token then s_2 and s_5 are empty. So, while some process is writing, no other process reads the buffer. s_2 carries at most k tokens: there are at most k processes reading concurrently. When no process is writing, that is, $M(s_4) = 0$, then s_2 may in fact obtain k tokens. Then the synchronization place s_5 is empty.

In particular, we prove the following

Proposition. *With the capacity K_N, defined as $K_N(s_i) = n$ for $i \in \{0, 1, 3\}$, $K_N(s_4) = 1$ and $K_N(s_2) = K_N(s_5) = k$, and with the initial marking M_N given in Fig. 67, the net shown in Fig. 66 is live.*

Proof. For the reasons discussed above the given capacity K_N will never hinder any firing of transitions. We start by showing that each marking $M \in [M_N\rangle$ enables at least one transition. In the case $M(s_0) + M(s_2) + M(s_4) > 0$, we see from the net structure that at least one of the transitions t_0, t_3, t_2 or t_5 is enabled. If $M(s_0)+M(s_2)+M(s_4)=0$, we get from i_1 that $M(s_1)+M(s_3)=n$, and from i_2 that $M(s_5)=k$. Then t_1 or t_4 is enabled. Now, if s_0 is empty for some $M \in [M_N\rangle$, it may be marked by some succession of firings. This implies the liveness of t_0 and t_3. The liveness of the other transitions follows immediately. □

6.4 Properties of a Sender-Receiver Model

As a modification of the producer-consumer model (Fig. 5 and Fig. 53), we discuss here a model consisting of a sender and a receiver. Both may terminate their activities by reaching a terminal state. The solution shown in Fig. 68 is not satisfactory since the receiver may reach its terminal state while the sender is not in its terminal state or the channel is not yet empty. To exclude these possibilities, we introduce a second channel (Fig. 69) which may carry a "terminated" message from the sender; additionally, the channel is supplemented by its complement allowing the possibility of testing whether it is empty.

This sender-receiver system is embedded into an environment, as represented in Fig. 70, which controls its activities. When the sender and the receiver reach their inactive state, they signal this to the environment. Then both may be restarted.

If the sender-receiver system is modelled correctly, it has the following properties:

(P_1) In each constellation, the sender is "inactive", "ready to send" or has just "finished sending". The receiver is "inactive", "ready to receive" or has just "finished receiving".

(P_2) The message channel contains at most n messages.

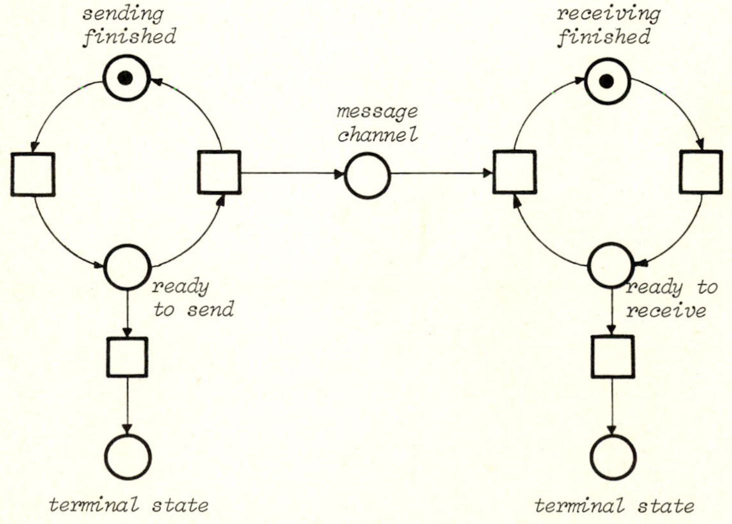

Fig. 68. Unsatisfactory version of a sender-receiver system with final states

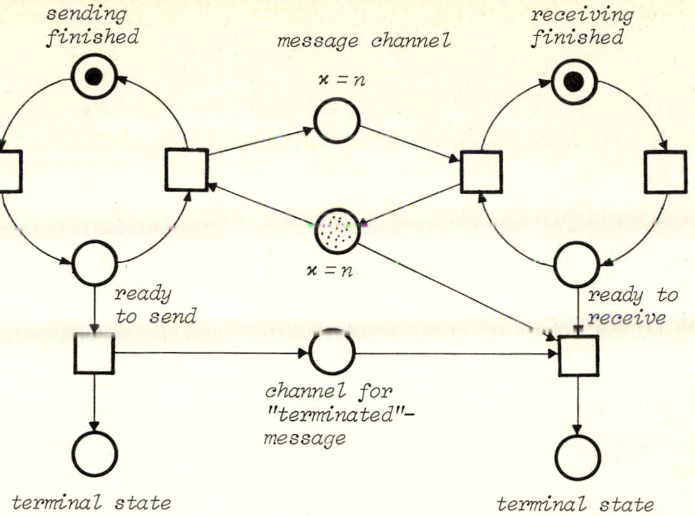

Fig. 69. Sender-receiver system with final states

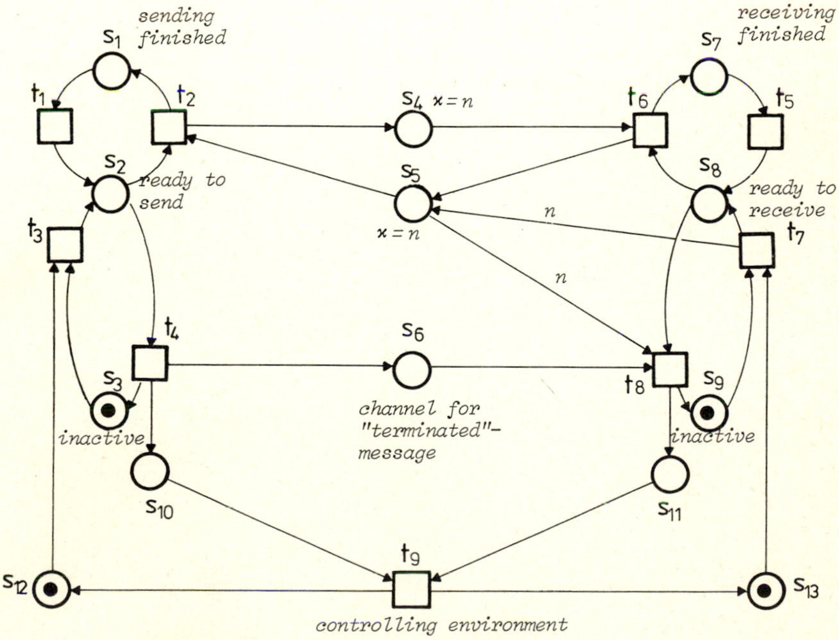

Fig. 70. The sender-receiver system, enlarged by a cyclic control

(P_3) The sender (and receiver, respectively) is inactive if and only if it sent a corresponding signal to the environment. It can leave the inactive state only as a result of a signal from the environment.

(P_4) If the sender has reached the inactive state, it cannot leave it again until the receiver has also reached its inactive state.

(P_5) The decision of the receiver whether to receive or whether to become inactive depends on the behaviour of the sender. In this respect, no conflict arises.

(P_6) The receiver may only become inactive if the channel is empty and the sender is inactive.

We prove these properties using the *S*-invariants shown in Fig. 71.

Let $M \in [M_N\rangle$ be an arbitrary reachable marking of M_N. Using i_1, we find $M(s_1) + M(s_2) + M(s_3) = 1$. Similarly using i_2: $M(s_7) + M(s_8) + M(s_9) = 1$. This proves ($P_1$).

	t_1	t_2	t_3	t_4	t_5	t_6	t_7	t_8	t_9	i_1	i_2	i_3	i_4	i_5	i_6	M_N
s_1	-1	1								1						
s_2	1	-1	1	-1						1						
s_3			-1	1						1			-1			1
s_4		1				-1						1				
s_5		-1				1	n	$-n$				1				
s_6				1				-1							1	
s_7					-1	1					1					
s_8					1	-1	1	-1			1					
s_9							-1	1			1	n		-1		1
s_{10}				1					-1				1		-1	
s_{11}								1	-1					1	1	
s_{12}			-1						1				1			1
s_{13}							-1		1					1		1

Fig. 71. Matrix, invariants $i_1, \ldots, i_6$ and the initial marking M_N of the net shown in Fig. 70

i_3 shows that the channel is correctly controlled, including the prevention of overflow: $M(s_4) + M(s_5) + n \cdot M(s_9) = n$. This implies ($P_2$) and, additionally, that the channel s_4 and its complement s_5 are both empty if and only if s_9 carries a token, that is, the receiver is inactive.

Property (P_3) for the sender follows from i_4 with $M(s_{10}) + M(s_{12}) - M(s_3) = 0$. This means that s_3 is marked if and only if s_{10} or s_{12} is marked. For the receiver, (P_3) follows in the same way from i_5. Using i_6, we have $M(s_6) - M(s_{10}) + M(s_{11}) = 0$. Hence $M(s_6) = 1$ implies $M(s_{10}) = 1$. This implies (P_4).

To show (P_5), we assume that t_6 and t_8 are both enabled by a marking $M \in [M_N\rangle$. Then, in particular, $M(s_4) \geq 1 \wedge M(s_5) \geq n \wedge M(s_8) \geq 1$, hence $M(s_4) + M(s_5) + M(s_8) \geq n + 2$. But, using the invariant $i_2 + i_3$, we have $M(s_4) + M(s_5) + M(s_7) + M(s_8) + (n+1) \cdot M(s_9) = n + 1$, hence $M(s_4) + M(s_5) + M(s_8) \leq n + 1$.

For (P_6): The receiver can reach the inactive state only when t_8 is enabled, that is, when $M(s_5) \geq n \wedge M(s_6) \geq 1 \wedge M(s_8) \geq 1$.

For such markings M, it has to be shown that

(1) $M(s_4) = 0$ and (2) $M(s_3) \geq 1$.

Suppose t_8 is enabled. So $M(s_5) \geq n$ and $M(s_6) \geq 1$.

(1) By i_3, $M(s_4) + M(s_5) + n \cdot M(s_9) = n$. So, $M(s_4) \leq 0$ (since $M(s_5) \geq n$ and $M(s_9) \geq 0$).

(2) $i_4 + i_6$ gives: $M(s_6) + M(s_{12}) + M(s_{11}) - M(s_3) = 0$. So $M(s_3) \geq M(s_6) \geq 1$.

6.5 A Seat-Reservation System

The stepwise development of a seat-reservation system is intended to show how models for planned systems can be constructed as P/T-nets. First, the system is represented as a net with inscriptions in English. Then it will be refined so that its structure corresponds to a P/T-net and its behaviour to the firing rule. By means of S-invariants, we shall prove some properties of the model.

Specification of the system: A seat-reservation system organizes the reservation of limited resources, for example the reservation of seats in aeroplanes. Several independent agencies (travel agencies) may access the system in order to book a seat or to cancel a reservation. In the case of a booking transaction, the system adds the customer to the passenger list; if the passenger list is full, he is added to a waiting list. In the case of a cancelling transaction, the customer is deleted from the passenger list or the waiting list, respectively. In each case, the customer gets a message; particularly, if the task may not be executed, for instance, in the case of repeated booking by the same customer or the cancellation of a reservation which has not been previously booked. The manager of the system may, using an updating routine, reserve released seats for customers on the waiting list and send them a message or, if the waiting list is empty, release those seats for direct reservation. Figure 72 shows the global view of the system.

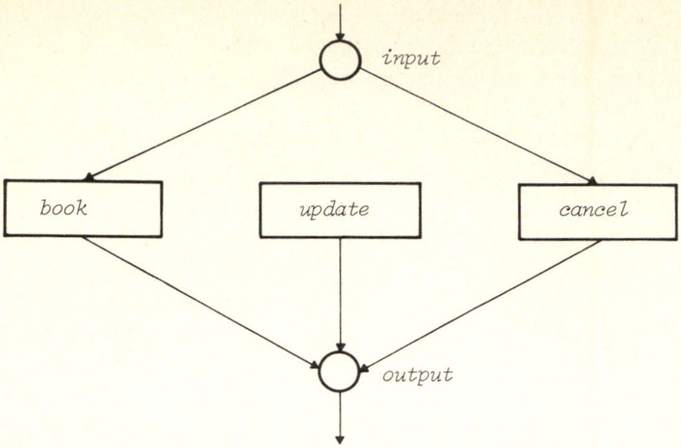

Fig. 72. Global view of the seat-reservation system

To achieve a high throughput, the system should handle transitions concurrently, as much as possible. In particular, booking and cancelling tasks should not hinder each other.

We shall prove the following three properties:

(P_1) It is not possible to overbook the passenger list.
(P_2) A customer is only added to the waiting list if the passenger list is full.
(P_3) A customer not in the waiting list is only directly added to the passenger list if the waiting list is empty; customers on the waiting list are the first to be served when reservations are cancelled.

Figure 73 shows the system as an inscribed net. The tasks from the travel agencies enter the system through the input place. Each task contains a customer identification and the booking or cancelling order; it should be considered as a labelled token. The conditions written into places (for example, $a=b$ or $i \in W$) have to be fulfilled to allow the associated transitions to fire. As in the representations of algorithms in Chap. 1 (Figs. 11, 13, 14), the inscriptions on transitions denote instructions, which are executed when the transition fires. Between instructions, the symbol "&" denotes concurrent execution, ";" denotes, as usual, sequential execution. The lists W and P are organized following a first-in-first-out principle, whereby *first* (W) denotes the first element of W.

An instruction $X \to W$ adds X to the end of the list W, *skip* (x, W) deletes x from the list W. $m_i := \ldots$ means an appropriate message is sent to the travel agency of customer i.

The instructions on one transition form an atomic action. This means that, during the execution of the instructions of some transition, the entities involved may not be changed by the firing of other transitions. (It would of course be possible to represent the organization of these indivisible executions by additional places in the net.) To achieve good performance, sections of indivisible instructions must be kept as small as possible. This is mainly

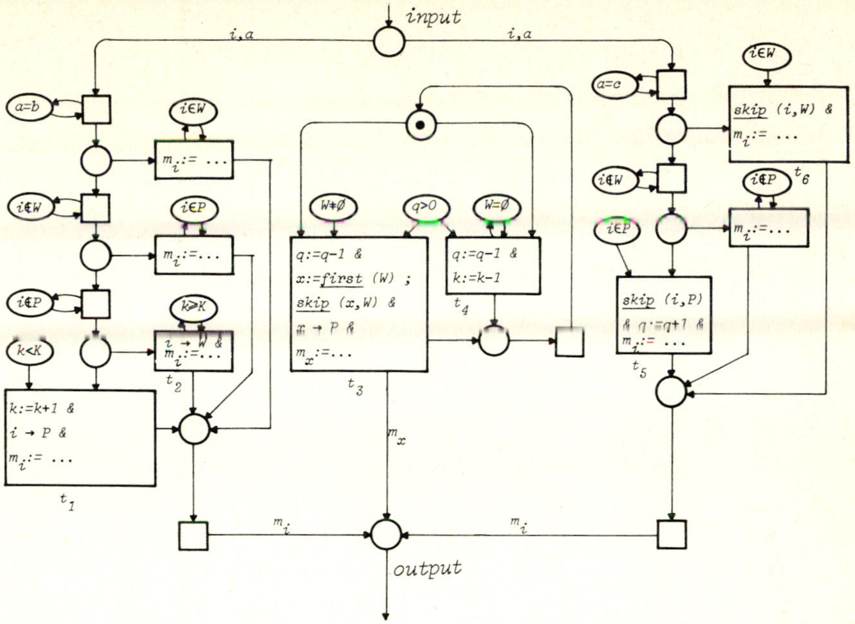

i:	*customer identification*
a:	*kind of order (b for booking or c for cancelling)*
m_i:	*message from the system to customer i*
W:	*waiting list*
P:	*passenger list*
K:	*capacity of P*
k:	*number of seats reserved in P*
q:	*number of cancelled reservations for which the seats are not yet released.*

Fig. 73. The seat-reservation system

achieved by the idea that cancelled seats are not immediately released for reservation again. Instead, they are counted by the variable q and may be processed by the updating module.

For considerations concerning liveness and boundedness, the dependencies between W, P, k and q are crucial. The influences from the environment can not be controlled within the system. Therefore, it is sufficient to consider the part of the system represented in Fig. 74 and to formalise these inscriptions. Thus, we have to presuppose that the six transitions $t_1, \dots, t_6$ are enabled at unforeseen intervals whenever the associated conditions are fulfilled. In particular, the messages to customers do not influence liveness and boundedness.

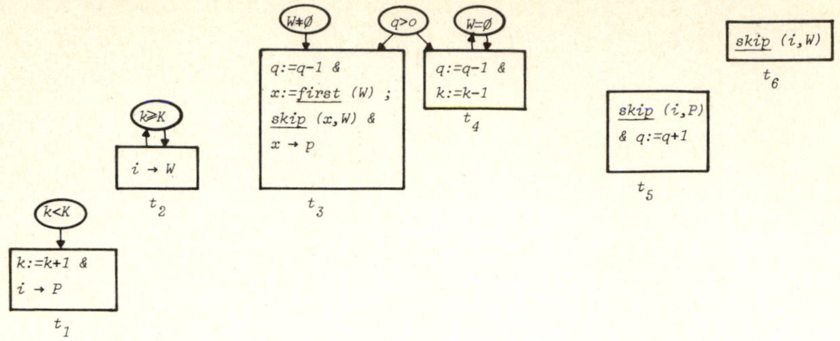

Fig. 74. The relevant part of Fig. 73 for correctness investigations

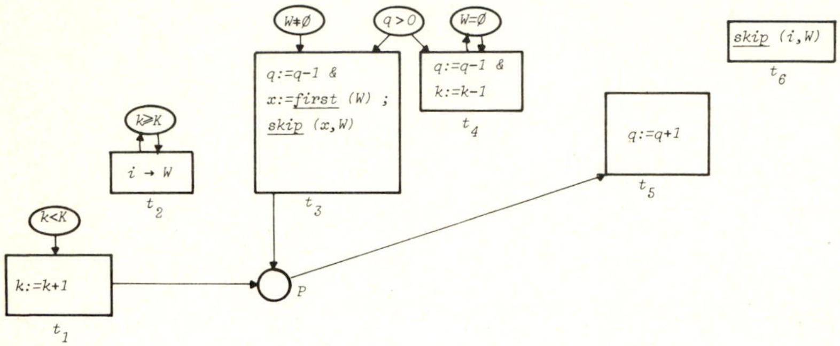

Fig. 75. Replacement of inscriptions concerning the passenger list in Fig. 74 by a new place P

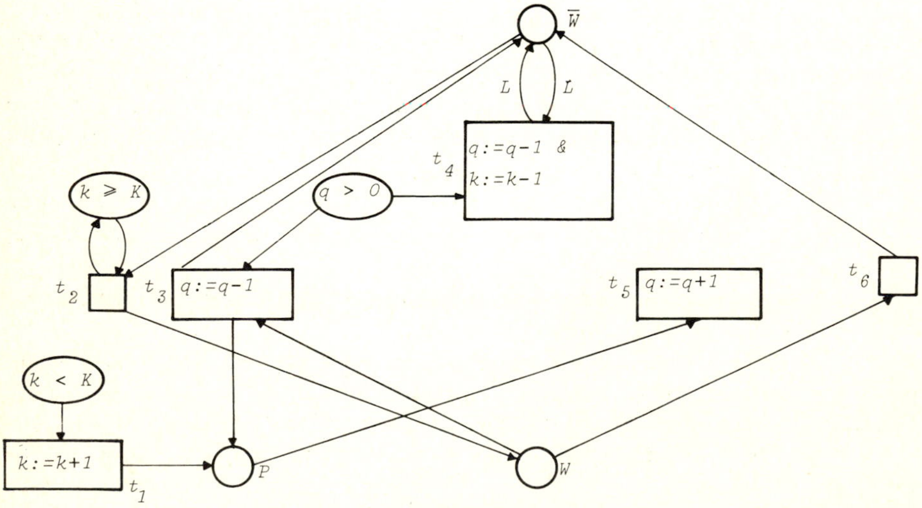

Fig. 76. Replacement of inscriptions concerning the waiting list in Fig. 75 by a place W and its complement $\bar{W}$

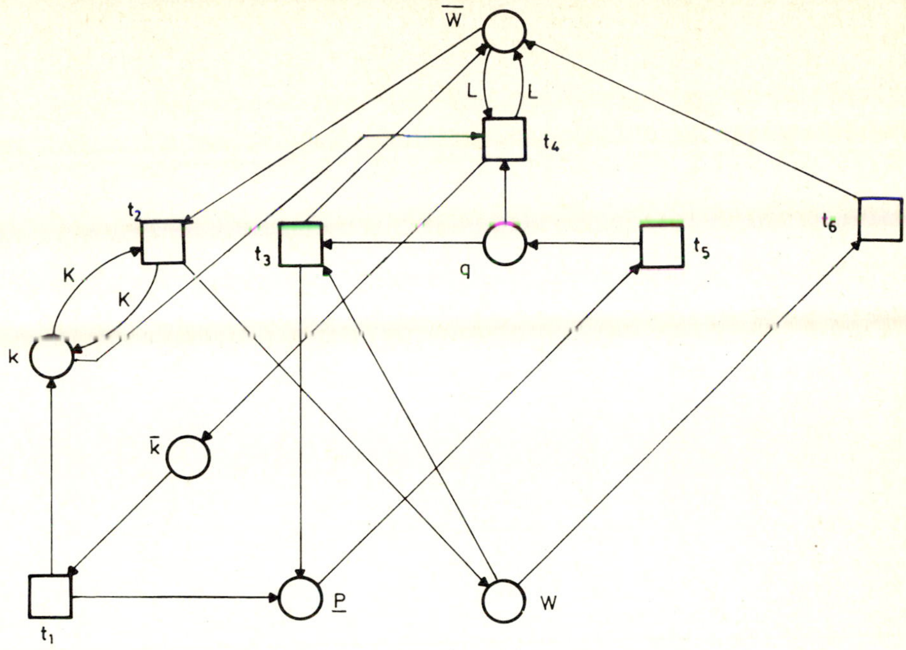

Fig. 77. Replacement of the remaining inscriptions of Fig. 76

To start with, we formalize the passenger list and its processing. To do this, a new place P is introduced and embedded in the system of Fig. 74 such that its token count represents the actual number of seats reserved in the passenger list. The corresponding inscriptions are deleted. Figure 75 shows the resulting system, whereby P is empty under the initial marking M_N.

As is the case of the passenger list, we organise the waiting list W as a new place W with $M_N(W) = 0$. Of course, the waiting list (as the passenger list) has a finite capacity, L. When it is also exhausted no further booking orders can be processed. As well as W, we also introduce the complementary place $\bar{W}$ with $M_N(\bar{W}) = L$. Figure 76 shows the result. (Notice that the introduction of complements $\bar{p}$ of places p serves to test emptyness of p.)

To replace the remaining inscriptions, we introduce places for q and k with $M_N(q) = M_N(k) = 0$ as shown in Fig. 77. For k, we also introduce the complement $\bar{k}$ with $M_N(\bar{k}) = K$.

The self-loops in the system of Fig. 77 are decomposed as shown in Fig. 78.

Using the invariants given in Fig. 79, we are now able to prove the properties (P_1), (P_2), (P_3) formulated above. In the following, let $M \in [M_N\rangle$ be an arbitrary reachable marking of M_N.

Using i_1, $M(P) + M(q) + M(\bar{k}) + M(y) = M_N(P) + M_N(q) + M_N(\bar{k}) + M_N(y) = K$. This implies $M(P) = K - M(q) - M(\bar{k}) - M(y) \leq K$ and hence (P_1).

Assume the passenger list P is totally booked. Then the number of actually reserved seats $M(P)$ together with the not yet released seats $M(q)$ exhausts

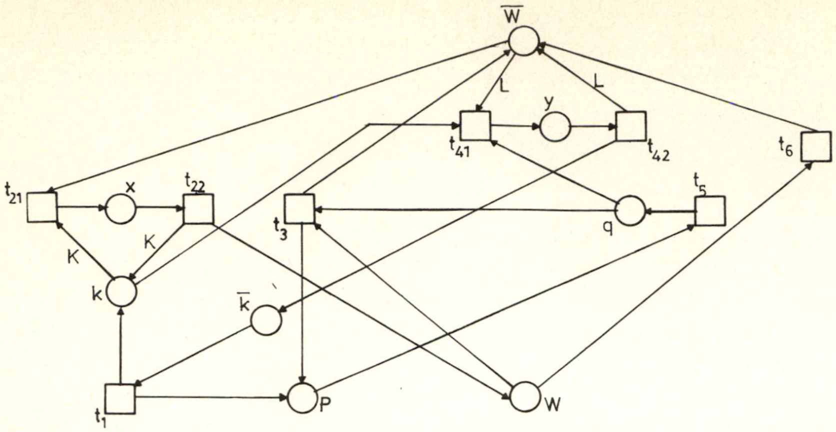

Fig. 78. Decomposition of self-loops in Fig. 77

the capacity K of P. x is marked if an only if the system answers a booking order by adding the customer to the waiting list. In this case we have, using i_2, $M(P)+M(q)-M(k)-K\cdot M(x)=0$ i.e. $M(P)+M(q)=M(k)+K\cdot M(X)$ $=M(k)+K\geq K$. This proves (P_2).

Let us now conversely assume a situation in which booking orders are answered by adding the customer to the waiting list. For the corresponding marking M_1 we have by (P_2): $M_1(k)=K$. Using i_3, $K\cdot M(X)+M(y)+M(k)$ $+M(\bar{k})=K$ and hence $M_1(\bar{k})=0$. Now, the updating module may release some seats for re-reservation and thus allow the system to react to a customer's

	t_1	t_{21}	t_{22}	t_3	t_{41}	t_{42}	t_5	t_6	i_1	i_2	i_3	i_4	M_N
x		1	−1							−K	K	1	
y					1	−1			1		1	L	
k	1	−K	K		−1					−1	1		
$\bar{k}$	−1					1			1		1		K
W			1	−1				−1				1	
$\bar{W}$		−1		1	−L	L		1				1	L
P	1			1			−1		1	1			
q				−1	−1		1		1	1			

Fig. 79. Matrix, invariants $i_1, \ldots, i_4$ and the initial marking M_N of the net shown in Fig. 78

booking by adding him to the passenger list (firing of t_1): $\bar{K}$ is marked. This is realised by firing t_{42} and requires that y was marked under some marking $M_2 \in [M_1\rangle$. Using i_4, $M(x) + L \cdot M(y) + M(W) + M(\bar{W}) = L$ and hence $M_2(W) = 0$. This proves (P_3).

6.6 The Verification of Facts in C/E-Systems by Means of S-Invariants

Since C/E-systems may be regarded as special P/T-nets, the invariant calculus is also applicable to them. In particular, it may be used for the verification of facts. Looking at the proof of Theorem 6.1 (d), we see immediately, for invariants i, that $M \cdot i = M_N \cdot i$ holds for all markings M which are reachable from M_N by forward and backward reasoning. This means for a C/E-system Σ, all $d, d' \in C_\Sigma$ and an invariant i, that $c_d \cdot i = c_{d'} \cdot i$ (again c_d denotes the characteristic vector of d, see A20). If i itself is a characteristic vector of some set of conditions $B \subseteq B_\Sigma$, $i = c_B$, then $c_d \cdot c_B = |d \cap B|$.

Consider again the two systems shown in Fig. 51 and Fig. 52. We shall show that the T-elements t and t_1, t_2, respectively, are facts by regarding these systems as P/T-nets with capacity one. The initial markings are the cases represented in Fig. 51 and Fig. 52 respectively.

First, we consider the system of Fig. 51. c_B, $B = \{b_1, \ldots, b_4\}$, is an S-invariant and we have, for the represented case d, $|d \cap B| = 1$, i.e. $c_d \cdot c_B = 1$. Using Theorem 6.1 (d), we have, for all reachable markings d', $c_{d'} \cdot c_B = 1$, i.e. $|d' \cap B| = 1$. Since $|{}^\bullet t \cap B| = 2$, t will never be enabled.

	e_1	e_2	e_3	e_4	i	d
b_1	-1			1		
b_2	1	-1				1
b_3		1	-1			
b_4			1	-1		
b_5		-1		1	1	1
b_6		-1	1		-1	1
b_7			-1	1	-1	1

Fig. 80. Matrix, an invariant and the initial marking of the net shown in Fig. 52

Figure 80 shows the matrix, an invariant i and the initial marking d of the system of Fig. 52. This yields $d \cdot i = -1$. Using i, we find, for all reachable markings M, $M(b_5) - M(b_6) - M(b_7) = -1$ and hence $M(b_6) + M(b_7) = M(b_5) + 1$. So, if b_6 and b_7 are marked then b_5 is also marked and t_2 is a fact. On the other hand, if b_5 is marked then, in particular, b_7 is also marked and t_1 is a fact.

There is no general rule how invariants can be applied for the verification of facts. How they can be applied depends on the particular case.

6.7 *T*-Invariants

In this section, we are concerned with solutions of systems of equations of the form $\underline{N} \cdot x = 0$. Let $v : T_N \to \mathbb{N}$ be such a solution. If it is possible, starting from some marking M, to fire each transition t exactly $v(t)$ times, then this again yields the marking M. This is explained by the following argument.

If c_t is the characteristic vector of $\{t\}$, $t \in T_N$, then $\underline{t} = \underline{N} \cdot c_t$. If $M_0 [t\rangle M_1$ then $M_0 + \underline{t} = M_1$ (Corollary 5.2 (b)) and hence $M_0 + \underline{N} \cdot c_t = M_1$. If $M_0 [t_1\rangle M_1 [t_2\rangle M_2$ we have $M_0 + \underline{t_1} + \underline{t_2} = M_2$ and hence $M_0 + \underline{N} \cdot c_{t_1} + \underline{N} \cdot c_{t_2} = M_0 + \underline{N}(c_{t_1} + c_{t_2}) = M_2$. Generalising this, with $M_0 [t_1\rangle \ldots [t_n\rangle M_n$,

$$M_n = M_0 + \sum_{i=1}^{n} \underline{t_i} = M_0 + \sum_{i=1}^{n} \underline{N} \cdot c_{t_i} = M_0 + \underline{N} \cdot \sum_{i=1}^{n} c_{t_i}.$$

We formalize these considerations in the following way:

(a) Theorem. *Let N be a P/T-net, let $M_0, \ldots, M_n \in [M_N\rangle$ and let $t_1, \ldots, t_n \in T_N$ such that $M_0 [t_1\rangle M_1 \ldots [t_n\rangle M_N$. Let $v : T_N \to \mathbb{N}$ be given by $v(t) = |\{i \mid 1 \leq i \leq n \wedge t_i = t\}|$. Then $M_0 + \underline{N} \cdot v = M_n$.*

Proof. By induction on n. $n = 0$: $M_0 + \underline{N} \cdot 0 = M_0 + 0 = M_0$. Now assume the proposition is true for $n-1$. For v': $T_N \to \mathbb{N}$, defined as $v'(t) = |\{i \mid 1 \leq i \leq n-1 \wedge t_i = t\}|$ we have by the induction hypothesis $M_0 + \underline{N} \cdot v' = M_{n-1}$. Furthermore $M_n = M_{n-1} + \underline{t_n} = M_0 + \underline{N} \cdot v' + t_n = M_0 + \underline{N} \cdot v' + \underline{N} \cdot c_{t_n} = M_0 + \underline{N}(v' + c_{t_n}) = M_0 + \underline{N} \cdot v$. □

The converse of this theorem is in general not true since, for the realization of some vector v: $T_N \to \mathbb{N}$, enough tokens and enough free capacities are needed.

(b) Theorem. *Let N be an unbounded P/T-net. Let $M, M' : S_N \to \mathbb{Z}$ and let v: $T_N \to \mathbb{N}$. Then $M + \underline{N} \cdot v = M'$ iff $\exists M'' : S_N \to \mathbb{N}\ \exists t_1, \ldots, t_n \in T_n$ such that $(M + M'') [t_1\rangle \ldots [t_n\rangle (M' + M'')$ and $\forall t \in T_N : v(t) = |\{i \mid 1 \leq i \leq n \wedge t_i = t\}|$.*

Proof. "$\Leftarrow$" Theorem 6.7 (a).

"⇒" Induction on $k = \sum_{i=1}^{n} v(t_i)$. $k = 0$: Then $M = M'$. The result follows with arbitrary M'' since $M + M'' [\emptyset\rangle M' + M''$.

Now assume that the proposition is true for $k-1$. Let $t \in T_N$ such that $v = v' + c_t$. Then $\sum_{i=1}^{n} v'(t_i) = k-1$. We have $M + \underline{N} \cdot v = M'$. Now let $M''' = M' - \underline{t}$. Then $M + \underline{N} \cdot v' = M + \underline{N} \cdot (v - c_t) = M + \underline{N} \cdot v - \underline{N} \cdot c_t = M + \underline{N} \cdot v - \underline{t} = M' - \underline{t} = M'''$.

By the induction hypothesis, there exists $n \in \mathbb{N}$, some marking M'', and $t_1, \ldots, t_n \in T_n$ such that $(M + M'') [t_1\rangle \ldots [t_n\rangle (M''' + M'')$, where $v'(t) = |\{i \mid 1 \le i \le n \wedge t_i = t\}|$. Now let $\tilde{M}: S_N \to \mathbb{N}$ be given by

$$\tilde{M}(s) = \begin{cases} W_N(s, t) & \text{if } s \in {}^{\bullet}t \\ 0 & \text{otherwise.} \end{cases}$$

Then $\tilde{M} [t\rangle \tilde{M} + \underline{t}$ and $(M + M'' + \tilde{M}) [t_1\rangle \ldots [t_n\rangle (M''' + M'' + \tilde{M}) [t_{n+1}\rangle (M''' + \underline{t} + M'' + \tilde{M})$, $t_{n+1} = t$, since $\forall s \in S_N : K_N(s) = \omega$. $M''' + \underline{t} = M'$ and $\forall t \in T_N : v(t) = |\{i \mid 1 \le i \le n+1 \wedge t_i = t\}|$. □

We are now able to investigate the relation between solutions of $\underline{N} \cdot x = 0$ and reproducible markings.

(c) Definition. A marking M of a P/T-net N is called *reproducible* iff $\exists M' \in [M\rangle$ with $M' \neq M$ and $M \in [M'\rangle$.

(d) Proposition. *Let N be a P/T-net with $\forall s \in S_N : K_N(s) = \omega$. If M is a reproducible marking and M' is an arbitrary marking of N then $M + M'$ is reproducible.*

(e) Definition. Let N be a P/T-net. A vector $i: T_N \to \mathbb{Z}$ is called a *T-invariant* iff $\underline{N} \cdot i = 0$.

(f) Corollary. *If i_1 and i_2 are T-invariants of a P/T-net N and $z \in \mathbb{Z}$ then $i_1 + i_2$ and $z \cdot i_1$ are also T-invariants.*

(g) Theorem. *Let N be a P/T-net with $\forall s \in S : K_N(s) = \omega$. N possesses a positive T-invariant $v \neq 0$ if and only if N possesses a reproducible marking.*

Proof. $\underline{N} \cdot v = 0 \Leftrightarrow 0 + \underline{N} \cdot v = 0 \Leftrightarrow \exists M''\ \exists t_1, \ldots, t_n \in T_N$ such that $(0 + M'') [t_1\rangle \ldots [t_n\rangle (0 + M'')$ and $\forall t \in T_N : v(t) = |\{i \mid 1 \le i \le n \wedge t_i = t\}|$ (Theorem 6.7 (b)). □

(h) Definition. A T-invariant i of a P/T-net N is called *realizable* iff there exists an $M_0 \in [M_N\rangle$ and a firing sequence $M_0 [t_1\rangle \ldots [t_n\rangle M_n$ such that $\forall t \in T_N : i(t) = |\{j \mid 1 \le j \le n \wedge t_j = t\}|$.

Not every positive T-invariant i of some P/T-net N is realizable; even if N is live and bounded and each marking of N is reproducible and i is not the sum of other positive invariants.

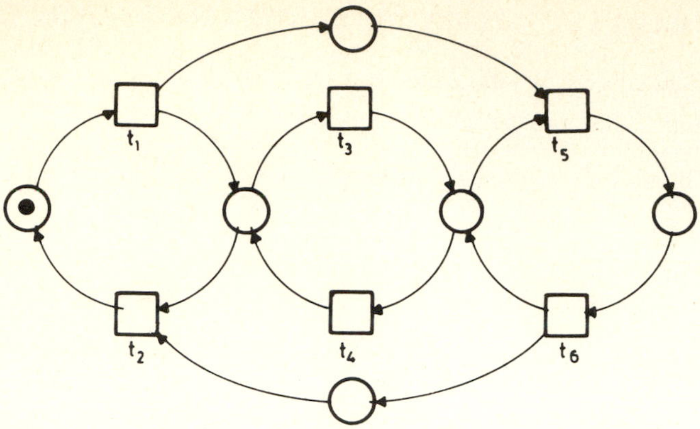

Fig. 81. The T-invariant i, given by $i(t_1) = i(t_2) = i(t_5) = i(t_6) = 1$ and $i(t_3) = i(t_4) = 0$, is not realizable

Figure 81 presents an example.

To conclude this section, we show that live and bounded P/T-nets are covered by T-invariants.

(i) Definition. A P/T-net is called *covered by T-invariants* iff, for each transition $t \in T_N$, there exists a positive T-invariant i of N with $i(t) > 0$.

(j) Corollary. *If a P/T-net N is covered by T-invariants then there exists a T-invariant i of N such that* $\forall t \in T_N : i(t) > 0$.

Proof. For $t \in T_N$, let i_t be a positive T-invariant with $i_t(t) > 0$. Then, using Corollary 6.7 (f), $i = \sum_{t \in T_N} i_t$ is a T-invariant fulfilling the requirements. □

(k) Theorem. *Every P/T-net which is finite, live and bounded is covered by T-invariants.*

Proof. If N is finite and live then $\forall M \in [M_N\rangle \ \exists \tilde{M} \in [M\rangle : M_0 [t_1\rangle \ldots [t_n\rangle M_n$ with $M_0 = M$ and $M_n = \tilde{M}$ and $T_N = \{t_1, \ldots, t_n\}$. If, furthermore, N is bounded, then $q = |[M_N\rangle| \in \mathbb{N}$. Then, for $i = 0, \ldots, q$, there exist firing sequences $M_i [t_{i_1}\rangle \ldots [t_{i_{n_i}}\rangle \tilde{M}_i$ with $T_N = \{t_{i_1}, \ldots, t_{i_{n_i}}\}$, $M_0 = M_N$ and $\tilde{M}_i = M_{i+1}$. Then there exist two indices $0 \leq j < k \leq q$ such that $M_j = M_k$ and a firing sequence $M_j [t'_1\rangle \ldots [t'_m\rangle M_k$ such that $\forall t \in T_N \ \exists 1 \leq i \leq m : t'_i = t$. Let the vector $v: T_N \to \mathbb{N}$ be defined as $v(t) = |\{i \mid 1 \leq i \leq m \wedge t'_i = t\}|$. Using Theorem 6.7 (a), $M_j + \underline{N} \cdot v = M_k$ and therefore $\underline{N} \cdot v = 0$, because $M_j = M_k$. Since $\forall t \in T_N : v(t) > 0$, v is a T-invariant which covers N. □

Exercises for Chapter 6

1. a) Compute some S-invariants of the P/T-net in Fig. 12.
   b) Is this net covered by S-invariants?

2. Show that the net in Exercise 6, Chap. 5, is not covered by S-invariants.

3. Show that the following net has T-invariants which are not realizable:

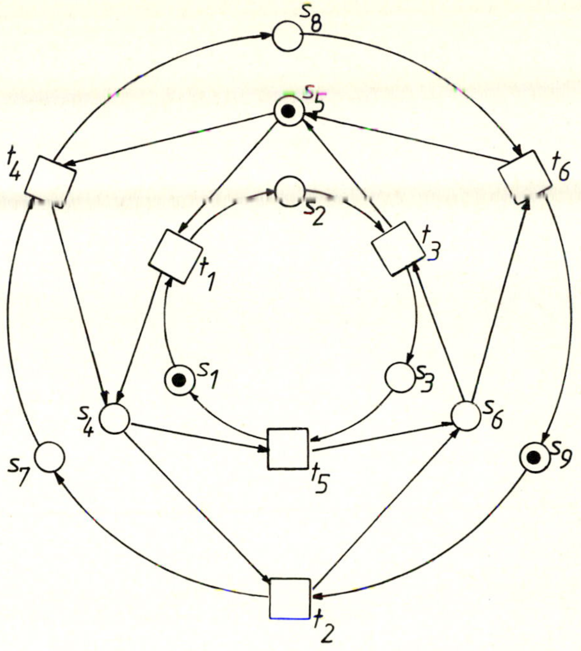

Chapter 7

Liveness Criteria for Special Classes of Nets

In this chapter, we investigate *marked nets;* these are special P/T-nets which are suitable for many applications. The liveness analysis for such nets is not much simpler than for P/T-nets in general, but there are special classes of marked nets for which criteria for liveness or safeness are known. These criteria are the main topic of this chapter.

7.1 Marked Nets, Deadlocks and Traps

(a) Definition. A P/T-net is called a *marked net* iff, for all $s \in S_N$, $M_N(s) \in \mathbb{N}$, $K_N(s) = \omega$, and for all $p \in F_N$, $W_N(p) = 1$.

When investigating liveness it is important to consider parts of the net which will never be marked or which will never lose all their tokens. In this section we shall consider such parts of nets and in particular we shall consider those in which such situations are easily recognizable.

A set S of places will never be marked again, after losing all tokens, if and only if no transition which contains in its postset a place belonging to S may ever fire again. In particular, this is the case if all these transitions also contain a place belonging to S in their preset, that is, $\forall t \in T_N : t \in {}^{\bullet}S \Rightarrow t \in S^{\bullet}$ or, equivalently, ${}^{\bullet}S \subseteq S^{\bullet}$ (Fig. 82). A set of places which meets this condition is called a *deadlock*. A deadlock may be found using the following procedure: Let s_0 be a place which belongs to a deadlock, S, we want to construct. Then, as well as s_0, for all transitions $t \in {}^{\bullet}s_0$, at least one place $s_1 \in {}^{\bullet}t$ must belong to S; that is ${}^{\bullet}s_0 \subseteq S^{\bullet}$. Now we iterate this and always require, for new elements $s \in S$, that ${}^{\bullet}s \subseteq S^{\bullet}$. The iteration terminates whenever $\forall s \in S$: ${}^{\bullet}s \subseteq S^{\bullet}$, that is whenever ${}^{\bullet}S \subseteq S^{\bullet}$. Hence a deadlock has been found.

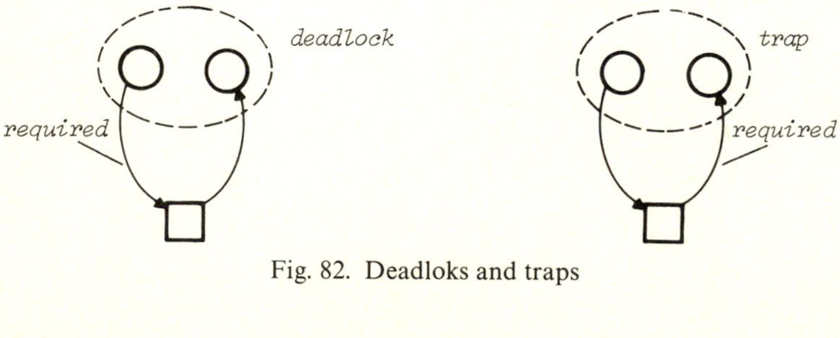

Fig. 82. Deadloks and traps

Deadlocks are critical system parts for liveness analysis, because transitions may never be enabled again if they contain places of an unmarked deadlock in their preset.

Dual to deadlocks, there are also system parts which will never lose all tokens again after they have once been marked. This is the case for some set of places, S, if every transition removing tokens from S also puts at least one token onto S. For this, we must have, for the set of transitions $S^\bullet$, that $S^\bullet \subseteq {}^\bullet S$ (Fig. 82). If S fulfils this condition then S is called a *trap*. A trap may be found using the following procedure: Let s_0 be a place which belongs to the trap, S, we want to construct. Then, as well as s_0, for all transitions $t \in s_0^\bullet$, at least one place $s_1 \in t^\bullet$ must belong to S; that is, $s_0^\bullet \subseteq {}^\bullet S$. Now we iterate this and always require, for new elements $s \in S$, that $s^\bullet \subseteq {}^\bullet S$, we terminate when $\forall s \in S$: $s^\bullet \subseteq {}^\bullet S$. This is equivalent to the condition $S^\bullet \subseteq {}^\bullet S$ derived above.

A deadlock which contains a marked trap as a subset will never become empty. Therefore, such deadlocks are important for liveness analysis.

(b) Definition. Let N be a marked net and let $S \subseteq S_N$.
(i) S is called a *deadlock* iff ${}^\bullet S \subseteq S^\bullet$.
(ii) S is called a *trap* iff $S^\bullet \subseteq {}^\bullet S$.

Examples of deadlocks and traps are shown in Fig. 83.

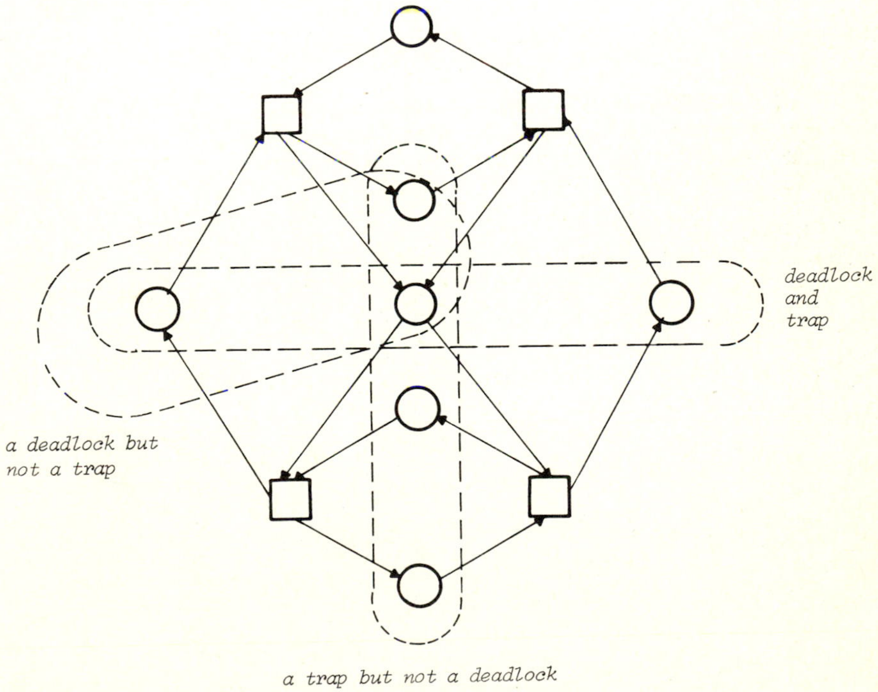

Fig. 83. Deadloks and traps

(c) Corollary. *Let N be a marked net with a positive S-invariant i and let $S = \{s \in S_N \mid i(s) > 0\}$. Then S is a deadlock and also a trap.*

Proof. This follows immediately from Corollary 6.1 (c). □

(d) Corollary. *Let N be a marked net, let $M\colon S_N \to \mathbb{N}$ be a marking of N and let $S \subseteq S_N$.*

(i) *If S is a deadlock which is unmarked under M then S is unmarked under each reachable marking $M' \in [M\rangle$.*
(ii) *If S is a trap which is marked under M then S is marked under each reachable marking $M' \in [M\rangle$.*
(iii) *The union of deadlocks is a deadlock.*
(iv) *The union of traps is a trap.*
(v) *S contains a maximal deadlock and a maximal trap.*

Proof. (i) Let S be unmarked under M, let $M[t\rangle M'$. Assume S is marked under M'. Then $t \in {}^{\bullet}S$. If S is a deadlock then $t \in S^{\bullet}$, but this is not possible since t is M-enabled.
(ii) Let S be marked under M, let $M[t\rangle M'$. Assume S is unmarked under M'. Then $t \in S^{\bullet}$. If S is a trap then $t \in {}^{\bullet}S$. So S is marked unter M'.
(iii) ${}^{\bullet}S_1 \subseteq S_1^{\bullet} \wedge {}^{\bullet}S_2 \subseteq S_2^{\bullet} \Rightarrow {}^{\bullet}(S_1 \cup S_2) = {}^{\bullet}S_1 \cup {}^{\bullet}S_2 \subseteq S_1^{\bullet} \cup S_2^{\bullet} = (S_1 \cup S_2)^{\bullet}$.
(iv) $S_1^{\bullet} \subseteq {}^{\bullet}S_1 \wedge S_2^{\bullet} \subseteq {}^{\bullet}S_2 \Rightarrow (S_1 \cup S_2)^{\bullet} = S_1^{\bullet} \cup S_2^{\bullet} \subseteq {}^{\bullet}S_1 \cup {}^{\bullet}S_2 = {}^{\bullet}(S_1 \cup S_2)$.
(v) follows using (iii) and (iv), since $\emptyset$ is a deadlock and a trap. □

For the class of all marked nets, we have the following relation between deadlocks, traps and reachable dead markings.

(e) Definition. Let N be a marked net and let $M\colon S_N \to \mathbb{N}$ be a marking of N. M is called *dead* iff no transition of T_N is M-enabled.

(f) Lemma. *Let N be a marked net. If $M\colon S_N \to \mathbb{N}$ is a dead marking then $S = \{s \in S_N \mid M(s) = 0\}$ is a non-empty, unmarked deadlock of N.*

Proof. Clearly $S \neq \emptyset$, otherwise all transitions would be M-enabled. S is a deadlock: Each transition $t \in {}^{\bullet}S$ is, by hypothesis, not M-enabled. Hence ${}^{\bullet}t \cap S \neq \emptyset$, i.e. $t \in S^{\bullet}$. By definition, S is unmarked. □

(g) Theorem. *Let N be a marked net. If each non-empty deadlock of N contains a trap which is marked under M_N then there is no dead marking in $[M_N\rangle$.*

Proof. Let $M \in [M_N\rangle$. Using Corollary 7.1 (d) (ii), each deadlock $S \neq \emptyset$ of N contains a trap which is marked under M. Hence each non-empty deadlock of N is marked under M. The Theorem follows from the above Lemma 7.1 (f). □

7.2 Free Choice Nets

In marked nets, a situation is called "confusion" (compare Sect. 2.1), if the enabeling of a transition t depends on the order in which two other transitions t', t'' fire. The analysis of liveness is particularly difficult in the presence of confusion. We shall now consider nets which exclude confusion by their structure, without regard of the marking class. A conflict between transitions $t_1, \ldots, t_n$ may only be resolved in favour of some transition t_i, $1 \leq i \leq n$; that is, t_i fires. This is achieved by the requirement that $t_1, \ldots, t_n$ possess only one common place $s \in {}^{\bullet}t_i$ and no further places in their presets. This means, in short, that the output transitions of a forward branched place may not be branched backwards. This is equivalent to the requirement that, for each arc $(s, t) \in F_N$, $s^{\bullet} = \{t\}$ or ${}^{\bullet}t = \{s\}$. Since, in such nets, one transition out of several transitions involved in a conflict may be chosen freely and independently to fire, they are called *free choice nets.*

(a) Definition. A marked net N is called a *free choice net* iff, for each arc $(s, t) \in F_N \cap (S_N \times T_N)$, $s^{\bullet} = \{t\} \vee {}^{\bullet}t = \{s\}$.

(b) Theorem. *The following properties of a marked net N are equivalent:*
(i) *N is a free choice net.*
(ii) $s \in S_N \wedge |s^{\bullet}| > 1 \Rightarrow \forall t \in s^{\bullet} : {}^{\bullet}t = \{s\}$.
(iii) $s_1, s_2 \in S_N \wedge s_1^{\bullet} \cap s_2^{\bullet} \neq \emptyset \Rightarrow \exists t \in T_N$ *with* $s_1^{\bullet} = s_2^{\bullet} = \{t\}$.
(iv) $s \in S_N \wedge |s^{\bullet}| > 1 \Rightarrow {}^{\bullet}(s^{\bullet}) = \{s\}$.

Proof. (i) $\Rightarrow$ (ii): If $|s^{\bullet}| > 1$ then, for each $t \in s^{\bullet}$, $s^{\bullet} \neq \{t\}$. Using (i), ${}^{\bullet}t = \{s\}$.
(ii) $\Rightarrow$ (i): Let $(s, t) \in F_N \cap (S_N \times T_N)$. If $|s^{\bullet}| = 1$ then immediately $s^{\bullet} = \{t\}$. If $|s^{\bullet}| > 1$, using (ii), ${}^{\bullet}t = \{s\}$.

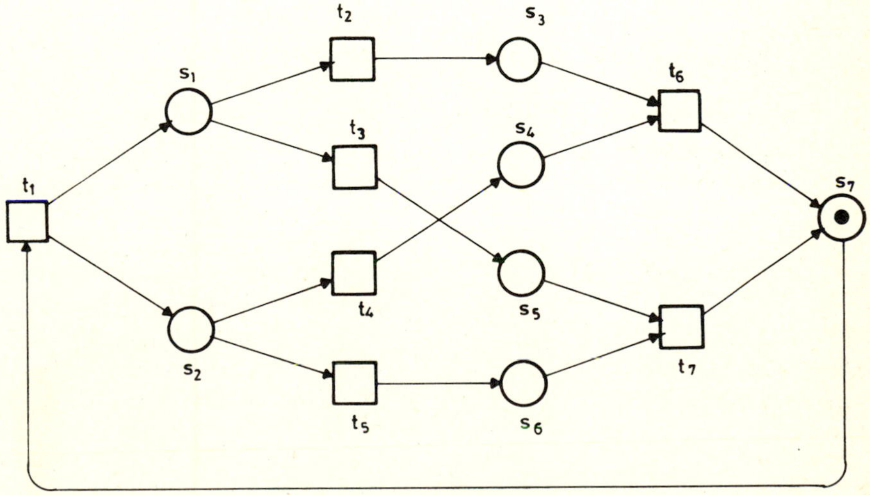

Fig. 84. A free choice net

(i) $\Rightarrow$ (iii): Let $t \in s_1^\bullet \cap s_2^\bullet$. Since $\{s_1, s_2\} \subseteq {}^\bullet t$, ${}^\bullet t \neq \{s_1\}$ and ${}^\bullet t \neq \{s_2\}$. Using (i), $s_1^\bullet = \{t\}$ and $s_2 = \{t\}$.

(iii) $\Rightarrow$ (i): Let $(s_1, t) \in F_N \cap (S_N \times T_N)$. If ${}^\bullet t \neq \{s_1\}$, there exists $s_2 \in S_N$, $s_2 \neq s_l$, with $t \in s_2^\bullet$. Then $t \in s_1^\bullet \cap s_2^\bullet \neq \emptyset$ and, using (iii), $s_1^\bullet = \{t\}$.

(iv) is obviously equivalent to (ii). □

The rest of this section is devoted to the derivation of a theorem, which states that a free choice net N is live if and only if each non-empty deadlock of N contains a trap which is marked under M_N. First we prove that this criterion is sufficient for liveness, and then that it is also necessary for liveness.

We start with some technical lemmas. Given a set of transitions, T, none of which may ever be enabled again, we show how to find further transitions of this kind.

(c) Lemma. *Let N be a free choice net and let $T \subseteq T_N$. If $({}^\bullet T)^\bullet$ may be enabled in $[M_N\rangle$ then T may be enabled in $[M_N\rangle$ too.*

Proof. Let $t_1 \in T$, $s \in {}^\bullet t_1$ and $t_2 \in s^\bullet \setminus T$ (Fig. 85). Since $t_1 \neq t_2$, we have $s^\bullet \neq \{t_1\}$ and $s^\bullet \neq \{t_2\}$. By the definition of free choice nets, ${}^\bullet t_1 = {}^\bullet t_2 = \{s\}$. t_2 is enabled if and only if s is marked. But, in this case, t_1 is enabled too.

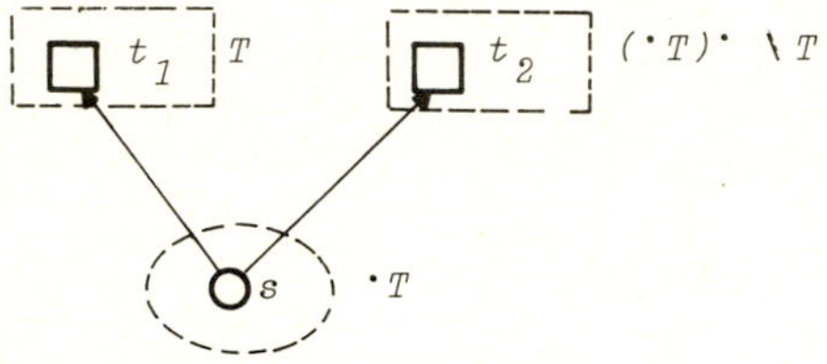

Fig. 85. Illustrating the proof of Lemma 7.2 (c)

□

(d) Definition. Let M be a marking of a net N. Then $\bar{M}$ denotes the set of places which are unmarked under M.

(e) Lemma. *Let N be a free choice net and let $T \subseteq T_N$ be a set of transitions none of which is enabled by any marking in $[M_N\rangle$. Then there exists a marking $M \in [M_N\rangle$ such that none of the transitions in ${}^\bullet({}^\bullet T \cap \bar{M}\rangle$ is enabled by any marking in $[M\rangle$.*

Proof. Let $M_0 \in [M_N\rangle$ be a marking such that there exists a transition $t \in {}^\bullet({}^\bullet T \cap \bar{M}_0)$ which fires to a marking M_1 and thereby marks a place $s \in {}^\bullet T \cap \bar{M}_0$ (Fig. 86). Using Lemma 7.2 (c), the transitions firing from M_N to M_1 do not belong to $({}^\bullet T)^\bullet$. Hence all places of ${}^\bullet T \setminus \bar{M}_N$ are marked under M_1 too, and therefore, in ${}^\bullet T$, only the places of ${}^\bullet T \cap \bar{M}_N$ are unmarked. Since s is marked under M_1, we have ${}^\bullet T \cap \bar{M}_1 \subsetneqq {}^\bullet T \cap \bar{M}_0 \subseteq {}^\bullet T \cap \bar{M}_N$.

By iterating this procedure (starting from M_1), we find in finitely many steps a marking M such that ${}^\bullet({}^\bullet T \cap \bar{M})$ may not be enabled in $[M\rangle$. Otherwise all elements of ${}^\bullet T$ could be marked. □

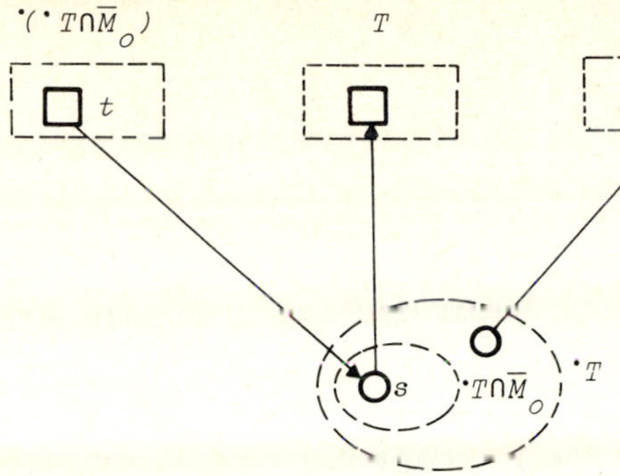

Fig. 86. Illustrating the proof of Lemma 7.2 (e)

We shall show indirectly that a free choice net N is live if every deadlock contains a trap marked under M_N. To do this we start from a set $T \subseteq T_N$ of transitions which may not be enabled in $[M_N\rangle$. We construct a deadlock $Q \subseteq {}^\bullet T$ which is unmarked under some reachable marking $M' \in [M_N\rangle$. Q contains traps (Corollary 7.1 (d) (v)). Using Corollary 7.1 (d) (ii), these traps must already be unmarked under M_N.

(f) Lemma. *Let N be a marked net and let $T \subseteq T_N$. If ${}^\bullet({}^\bullet T \cap \bar{M}_N) \subseteq T$ then either there exists a transition in T which is M_N-enabled or ${}^\bullet T \cap \bar{M}_N$ is an unmarked deadlock.*

Proof. Assume no transition in T is M_N-enabled. Let $Q = {}^\bullet T \cap \bar{M}_N$ and let $t \in {}^\bullet Q$. By the hypothesis, $t \in T$. Since T is not M_N-enabled, ${}^\bullet t \cap \bar{M}_N \neq \emptyset$ and hence ${}^\bullet t \cap Q \neq \emptyset$, that is $t \in Q^\bullet$ (Fig. 87). Since this is true for each $t \in {}^\bullet Q$ we have ${}^\bullet Q \subseteq Q^\bullet$.

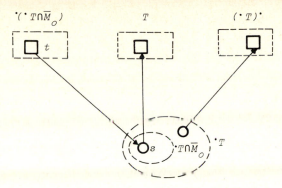

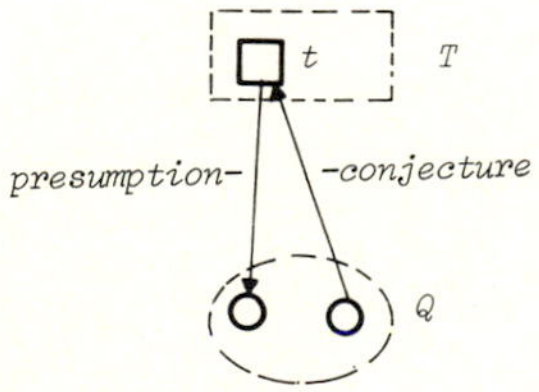

Fig. 87. Illustrating the proof of Lemma 7.2 (f)

□

(g) Theorem. *Let N be a free choice net and let $T \subseteq T_N$ be a set of transitions none of which is enabled by any marking in $[M_N\rangle$. Then there exists a marking $M \in [M_N\rangle$ and a deadlock of N which is unmarked under M.*

Proof. By induction on $|T_N \backslash T|$. $|T_N \backslash T| = 0$: Since $T_N = T$, trivially ${}^\bullet({}^\bullet T \cap \bar{M}_N) \subseteq T$. Using Lemma 7.2 (f), ${}^\bullet T \cap M_N$ is an unmarked deadlock. Induction hypothesis: The proposition is true if $|T_N \backslash T| = n$. Now let $|T_N \backslash T| = n+1$. Using

Lemma 7.2 (e), there exists a marking $M \in [M_N\rangle$ such that no transition in $^{\bullet}(^{\bullet}T \cap \bar{M})$ may be enabled in $[M\rangle$. If $^{\bullet}(^{\bullet}T \cap \bar{M}) \subseteq T$ the result follows using Lemma 7.2 (f). Otherwise, let $t \in {}^{\bullet}(^{\bullet}T \cap M)\backslash T$. Since $T \cup \{t\}$ may not be enabled in $[M\rangle$ (Lemma 7.2 (e)) and $|T_N \backslash (T \cup \{t\})| = n$, we have by the induction hypothesis: There exists a marking $M' \in [M\rangle$ such that some deadlock of N is unmarked under M'. In particular, $M' \in [M_N\rangle$. □

(h) Corollary. *Let N be a free choice net. If every non-empty deadlock contains a trap which is marked under M_N then N is live.*

Proof. If N is not live then there exists a marking $M \in [M_N\rangle$ and a non-empty set of transitions which may not be enabled in $[M\rangle$. Then, using Theorem 7.2 (g), there exists a marking $M' \in [M_N\rangle$ and a deadlock Q which is unmarked under M'. Corollary 7.1 (d) states that Q may not become empty in $[M_N\rangle$ if Q contains a trap which is marked under M_N. □

We have derived a criterion for the liveness of a free choice net and shown that it is a sufficient condition. Next, we shall show that it is also a necessary condition. For this, we assume a non-empty deadlock Q which does not contain a marked trap under the initial marking. By firing the appropriately chosen transitions of $Q^{\bullet}\backslash{}^{\bullet}Q$ the token count on Q is reduced until no transition of $Q^{\bullet}$ may fire any more. This is possible if all traps of Q are unmarked. Then only tokens of the places in $Q\backslash Q_1$, where Q_1 is the maximal trap in the deadlock Q, have to be removed as far as possible. To each place $s \in Q\backslash Q_1$, a transition $\alpha(s) \in s^{\bullet}$ is allocated. One difficulty is that these transitions $\alpha(s)$ have to be fired in such a way that those transitions $\alpha(s)$ which are not enabled may not be enabled again.

(i) Definition. Let N be a marked net and let $S \subseteq S_N$. A mapping $\alpha: S \rightarrow S^{\bullet}$ is called an *allocation.*

An allocation α is called *cycle-free* iff there is no set of places $\{s_0, \ldots, s_n\} \subseteq S$ such that $s_i \in \alpha(s_{i-1})^{\bullet}$ $(i = 1, \ldots, n)$ and $s_0 \in \alpha(s_n)^{\bullet}$. An allocation α partitions $S^{\bullet}$ into the set $\alpha(S)$ of images of α and the set $\bar{\alpha}(S) = S^{\bullet}\backslash\alpha(S)$.

(j) Lemma. *Let N be a marked net and let $S \subseteq S_N$ be an arbitrary set of places. Let $Q_1 \subseteq S$ be the maximal trap in S and let $Q_2 = S\backslash Q_1$. Then there exists a cycle-free allocation α: $Q_2 \rightarrow Q_2^{\bullet}$ such that $\alpha(Q_2) \cap {}^{\bullet}Q_1 = \emptyset$.*

Proof. By induction on $|Q_2|$. $|Q_2| = 0$: Then $\alpha: \emptyset \rightarrow \emptyset$ fulfils the requirements. Induction hypothesis: The proposition is true if $|Q_2| = n$. Now let $|Q_2| = n+1$. Then there exists some place $s_0 \in Q_2$ and some transition $t \in T_N$ such that $s_0 \in {}^{\bullet}t$ and $t^{\bullet} \cap Q_1 = \emptyset$ (Fig. 88). With $Q_2' = Q_2\backslash\{s_0\}$, Q_1 is the maximal trap in $Q_1 \cup Q_2'$. Then, by the induction hypothesis, there exists a cycle-free allocation α': $Q_2' \rightarrow Q_2'^{\bullet}$ such that $\alpha'(Q_2') \cap {}^{\bullet}Q_1 = \emptyset$. Now we define the allocation α: $Q_2 \rightarrow Q_2^{\bullet}$ by $\alpha(s) = \alpha'(s)$ for $s \in Q_2'$, $\alpha(s_0) = t$. Since $\alpha(s_0) \cap {}^{\bullet}Q_1 = t \cap {}^{\bullet}Q_1 = \emptyset$, we have $\alpha(Q_2) \cap {}^{\bullet}Q_1 = \alpha(Q_2' \cup \{s_0\}) \cap {}^{\bullet}Q_1 = \emptyset$. Since $\alpha(s_0)^{\bullet} \cap Q_2 = t^{\bullet} \cap Q_2 = \emptyset$,

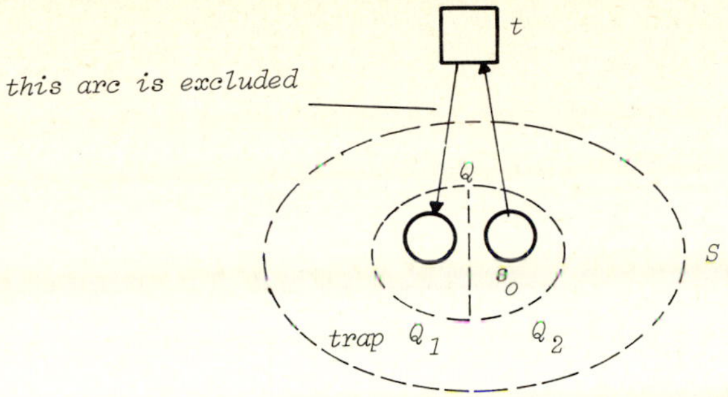

Fig. 88. Illustrating the proof of Lemma 7.2 (j)

s_0 does not belong to any cycle of Q_2. Therefore, as α' is cycle-free by the induction hypothesis, α is also cycle-free. □

Figure 89 shows an example illustrating this lemma.

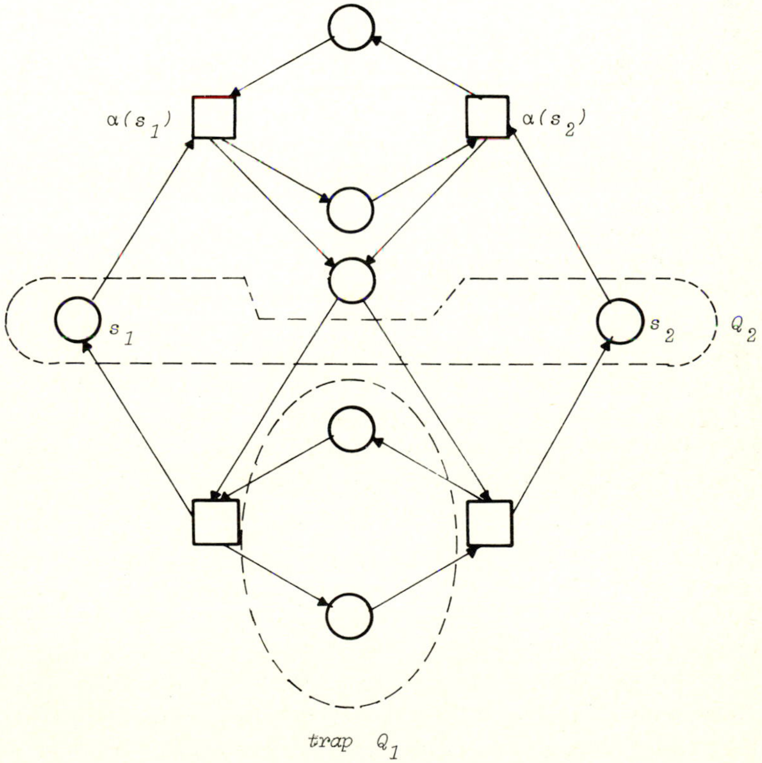

Fig. 89. An example for Lemma 7.2 (j)

(k) Theorem. *Let N be a free choice net and let $Q \subseteq S_N$ be a deadlock such that the maximal trap of Q is unmarked under M_N. Then there exists a marking $M \in [M_N\rangle$ such that $Q^{\bullet}$ may not be enabled in $[M\rangle$.*

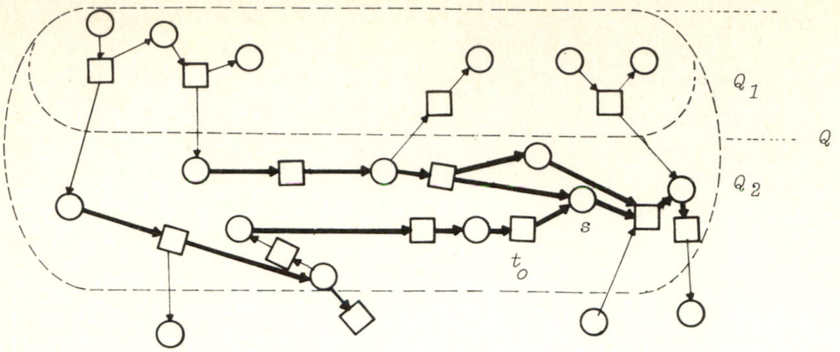

Fig. 90. A deadlock Q with the maximal trap Q_1. The thick arcs represent a cycle free allocation of Q_2. The sets $B_0, \ldots, B_3$ of places are also represented

Proof. Let Q_1 be the maximal trap of Q and let $Q_2 = Q \backslash Q_1$. Using Lemma 7.2 (j), there exists some cycle-free allocation α of Q_2 such that $\alpha(Q_2) \cap {}^{\bullet}Q_1 = \emptyset$.

The following notions will be applied during this proof: Call a marking $M_n \in [M_N\rangle$ *properly reached* iff there exists a firing sequence $M_N [t_1\rangle M_1 \ldots M_{n-1} [t_n\rangle M_n$ such that $\forall 1 \le i \le n$: $t_i \notin \bar{\alpha}(Q_2)$.

For $s, s' \in Q_2$ let $s < s'$ iff $\exists s_0, \ldots, s_n$ *with* $\alpha(s_i) \in {}^{\bullet}s_{i+1}$ $(i = 1, \ldots, n) \wedge s_0 = s \wedge s_n = s'$. $S \subseteq Q_2$ is *left closed* iff $\forall s \in S, s' \in Q_2: s' < s \Rightarrow s' \in S$.

A subset $S \subseteq Q_2$ is *detached by a marking M* iff for all properly reached markings $M' \in [M\rangle$, no transition in $\alpha(S)$ is M'-enabled. The proof will be based on five propositions:

Proposition 1. *For some place $s \in Q_2$ let $S = \{s' \in Q_2 | s' < s\}$ be detached by a marking M. Then there exists a properly reached marking $M' \in [M\rangle$ such that $S \cup \{s\}$ is detached by M'.*

Proof. As S is detached, no transition $t \in {}^{\bullet}s \cap \alpha(Q_2)$ can be fired by markings which are properly reached from M (as ${}^{\bullet}s \subseteq \alpha(Q_2) \cup \bar{\alpha}(Q_2)$). So, in the class of properly reached markings, $\alpha(s)$ can not be fired more than $M(s)$ times. Hence there exists some properly reached M' such that $S \cup \{s\}$ is detached by M'.

Proposition 2. *There exists a properly reached marking M such that Q_2 is detached by M.*

Proof. Let $Q_2 = \{s_1, \ldots, s_n\}$ such that, for all $0 \le i \le n$, $S_i = \{s_1, \ldots, s_i\}$ is left closed. As Q_2 is finite, this can easily be achieved.

For each subset S_i $(0 \le i \le n)$ we show by induction on i that there exists a properly reached marking $M_i \in [M_N\rangle$ such that S_i is detached by M_i. For $i = 0$, $S_i = \emptyset$ and the Proposition holds with $M = M_N$.

By induction hypothesis assume a properly reached marking $M_i \in [M_N\rangle$ such that S_i is detached by M_i. With Proposition 1, there exists a properly reached marking $M_{i+1} \in [M_i\rangle$ such that S_{i+1} is detached by M_{i+1}. Obviously, M_{i+1} is properly reached from M_N. For $i = n$, the Proposition follows.

Proposition 3. *If M is a properly reached marking, Q_1 is unmarked.*

Proof. By induction on the set of properly reached markings: By assumption of the Theorem, Q_1 is unmarked by M_N. Assume Q_1 to be unmarked by M and let $M[t\rangle M'$. We have to show that $t \notin {}^{\bullet}Q_1$. Note that ${}^{\bullet}Q_1 \subseteq {}^{\bullet}Q \subseteq Q^{\bullet} = Q_1^{\bullet} \cup Q_2^{\bullet} = Q_1^{\bullet} \cup \alpha(Q_2) \cup \bar{\alpha}(Q_2)$. Obviously $t \notin Q_1^{\bullet}$ as Q_1 is unmarked by M. $t \notin \bar{\alpha}(Q_2)$, if M' is to be properly reached. If $t \in \alpha(Q_2)$, by construction of α, $t \notin {}^{\bullet}Q_1$.

Proposition 4. *Let M be a properly reached marking such that Q_2 is detached by M. Then no transition $t \in Q^{\bullet}$ is M-enabled.*

Proof. By construction, $Q^{\bullet} = Q_1^{\bullet} \cup \alpha(Q_2) \cup \bar{\alpha}(Q_2)$. For $t \in Q_1^{\bullet}$, apply Proposition 3. For $t \in \alpha(Q_2)$ notice that Q_2 is detached by M. So, let $t \in \bar{\alpha}(Q_2)$. Then there exists a place $s \in Q_2$ such that $t \in s^{\bullet}$ and $t \neq \alpha(s)$. By the free choice properly of N, ${}^{\bullet}t = {}^{\bullet}\alpha(s) = s$. As $\alpha(s)$ is not M-enabled, $M(s) = 0$, hence t is also not M-enabled.

Proposition 5. *Let M be a properly reached marking such that Q_2 is detached by M. Then each marking $M' \in [M\rangle$ is properly reached and $Q_2^{\bullet}$ is detached by M'.*

Proof. By induction on the structure of $[M\rangle$. For M the Proposition holds by assumption. So, let $M' \in [M\rangle$ be properly reached and let Q_2 be detached by M'. For $M'[t\rangle M''$ we have to show that $t \notin \bar{\alpha}(Q_2)$. This follows from Proposition 4, as $\bar{\alpha}(Q_2) \subseteq Q^{\bullet}$.

We show the Theorem now as follows:

By Proposition 2, let $M \in [M_N\rangle$ be properly reached such that Q_2 is detached by M. By Proposition 5, each $M' \in [M_N\rangle$ is properly reached and Q_2 is detached by M'. The Theorem follows with Proposition 4. □

(l) Corollary. *A free choice net N is live if and only if every non-empty deadlock of N contains a trap which is marked under M_N.*

Proof. "⇐" Corollary 7.2 (h).

"⇒" Let Q be a deadlock such that all traps of Q are unmarked. Then the maximal trap of Q (the union of all traps of Q) is also unmarked and the result follows using the above theorem. □

Using this corollary we can easily verify that the marked net shown in Fig. 84 is not live, moreover that there is no initial marking under which it is live. Clearly, $Q = \{s_1, s_2, s_3, s_6, s_7\}$ is a deadlock, since $Q^{\bullet} = T_N \supseteq {}^{\bullet}Q$. But this deadlock does not contain any non-empty trap.

As an immediate consequence of this result we obtain that any enlargement of the initial marking of free choice nets preserves liveness.

(m) Corollary. *Let N and N' be free choice nets such that $(S_N, T_N; F_N) = (S_{N'}, T_{N'}; F_{N'})$ and $M_N \leq M_{N'}$. Then the liveness of N implies the liveness of N'.*

Proof. N is live $\Rightarrow$ each non-empty deadlock of N contains a trap which is marked under M_N (Corollary 7.2 (1)) $\Rightarrow$ each non-empty deadlock of N' contains a marked trap under $M_{N'}$ $\Rightarrow$ N' is live (Corollary 7.2 (1)). □

Figure 61 shows that this conjecture turns out to be false for the general case of marked nets.

7.3 Marked Graphs

To conclude this chapter, we investigate nets with only unbranched places. As in such nets every place possesses exactly one pre- and one post-transition, no conflict situations are possible. Such nets describe systems which are only structured by synchronization of their active elements. They are wellknown under the name *marked graph.*

Liveness and safeness of marked graphs are characterizable by very simple properties.

(a) Definition. A marked net is called a *marked graph* iff
(i) $\forall t_1, t_2 \in T_N : t_1 \, (F_N^*) \, t_2$ (N is strongly connected),
(ii) $\forall s \in S_N : |{}^\bullet s| = |s^\bullet| = 1$ (places are unbranched).

Examples of marked graphs are shown in Figs. 1, 2, 3, 21 and 42.

An important property of marked graphs is that the token count on each cycle does not change when transitions fire.

(b) Definition. Let N be a marked graph. A sequence $w = (s_0, \ldots, s_n)$ of places is called a *path of length n* iff, for $i = 1, \ldots, n$, $s_{i-1}^\bullet = {}^\bullet s_i$ and for all $1 \le i \ne j \le n$ $s_i^\bullet \ne s_j^\bullet \wedge {}^\bullet s_i \ne {}^\bullet s_j$. w *starts* at ${}^\bullet s_0$ and *ends* at $s_n^\bullet$. w is called a *cycle* iff w is a path such that ${}^\bullet s_0 = s_n^\bullet$.

(c) Lemma. *Let N be a marked graph and let* $(s_0, \ldots, s_n)$ *be a cycle of N. Then, for all markings* $M \in [M_N\rangle$, $\sum_{i=0}^{n} M(s_i) = \sum_{i=0}^{n} M_N(s_i)$.

Proof. Let $M_1 [t\rangle M_2$ denote a firing in N.

First case: $t = {}^\bullet s_i$ for some $0 \le i \le n$. The firing of t decreases the number of tokens on s_i by one and increases the number of tokens on s_{i+1} by one (let $s_{n+1} = s_0$). The marking of all other places of the cycle is not affected. Second case: $t \notin {}^\bullet\{s_0, \ldots, s_n\}$. The marking of all places belonging to the cycle remains unchanged. □

(d) Corollary. *If a set of places of a marked graph is a cycle then its characteristic vector is an S-invariant.*

Liveness of marked graphs may be characterized in a simple way:

(e) Theorem. *Let N be a marked graph. N is live if and only if every cycle of N contains at least one place which is marked under M_N.*

Proof. If there is a cycle which has all places unmarked under M_N then, using Lemma 7.3 (c), these places are also unmarked under all markings reachable from M_N. Hence the transitions belonging to this cycle may not be enabled in $[M_N\rangle$.

Conversely let $M \in [M_N\rangle$. Using Lemma 7.3 (c), every cycle contains at least one place which is marked under M. Since N is finite, there may not be arbitrary long paths in N such that all places on the path are unmarked under M.

Now let $t \in T_N$ and let n be the maximal length of the unmarked paths under M_N, ending with t. The start transition of each such path is enabled (otherwise there would be a longer unmarked path). Now it is possible to fire all these transitions independently of each other. This yields a marking $M \in [M_N\rangle$ such that the maximal length of the paths unmarked under M, ending at t, is $n-1$. The iteration of this procedure yields, after $n-1$ steps, a marking such that t is enabled. □

(f) Definition. A P/T-net N is called *safe* iff, for all $M \in [M_N\rangle$ and all $s \in S_N$, $M(s) \leq 1$.

(g) Theorem. *Let N be a marked graph which is live. N is safe if and only if each place $s \in S_N$ belongs to a cycle, which possesses exactly one place which is marked under M_N.*

Proof. By Lemma 7.3 (c), this condition is sufficient for safeness.

Now let $s \in S_N$ be a place, which belongs only to cycles which carry more that one token. Since N is live, the transition in $^{\bullet}s$ may be enabled and there exists a marking $M \in [M_N\rangle$ with $M(s) = 1$. Now we remove temporarily this token from s. By Theorem 7.3 (e) this does not affect liveness, since every cycle still possesses at least one marked place. Again the transition in $^{\bullet}s$ may be enabled. After its firing, s now contains two tokens, including the token which we removed temporarily. So N is not safe. □

(h) Corollary. *A marked graph N is live and safe if every cycle of N contains at least one marked place and if every place of N belongs to a cycle which contains exactly one marked place.*

Exercises for Chapter 7

1. For which initial markings M_N of the following net N do not any dead reachable markings $M \in [M_N\rangle$ exist?

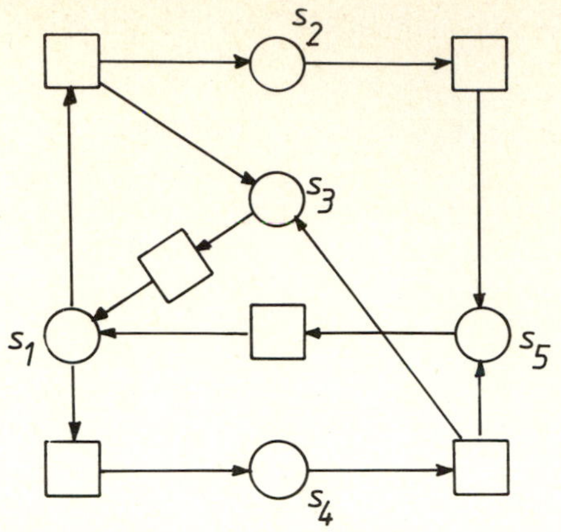

2. Does an initial marking exist such that the following net is live?

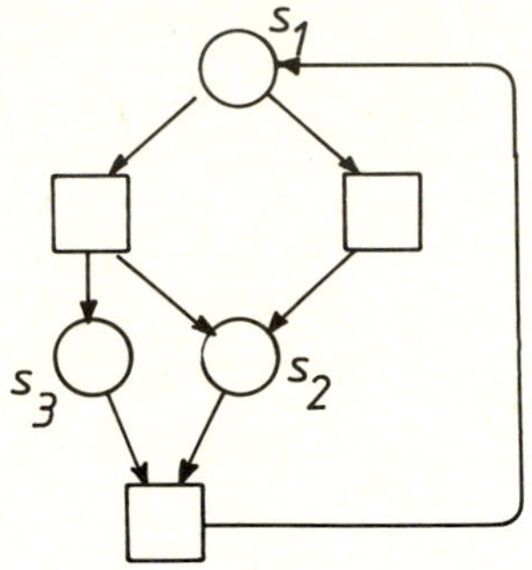

3. Construct an initial marking such that the following marked graph is live and safe:

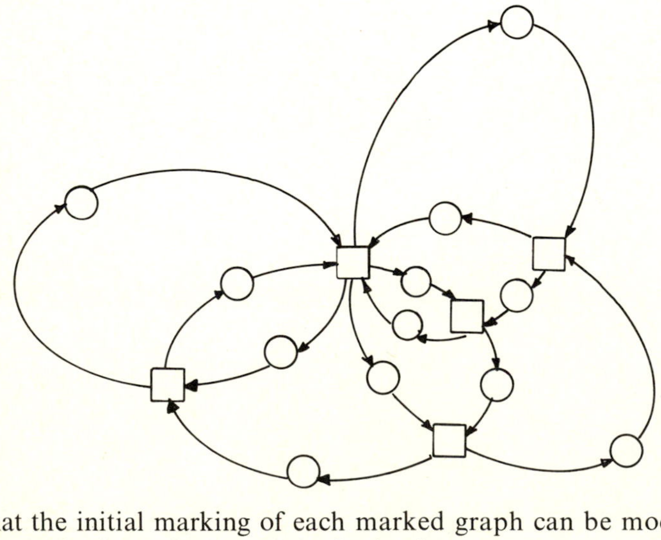

*4. Show that the initial marking of each marked graph can be modified such that a live and safe marked graph is obtained.

Part 3. Nets with Individual Tokens

The markings of the nets considered so far are fully determined by the number and the distribution of tokens on the *S*-elements. Now we shall allow individual objects as tokens. A marking then also depends on the nature of its tokens. We have already seen an example for such a net, called a *predicate/event-net,* with the library system in Fig. 18. As in Sect. 4.5, we shall show how relations between individuals which hold in all cases may be formulated in predicate logic. Again, they may be represented, in such nets as *T*-elements which are never enabled. A concept of "invariants" (as used for *P/T*-nets in Chap. 6) again helps us to verify properties of such nets. Such invariants will be defined for *relation nets,* which are introduced in Chap. 9.

The step from predicate/event-nets to relation nets is the same as from *C/E*-systems to *P/T*-nets: Instead of single individual objects we allow several individuals of the same kind. Then a linear algebraic calculus may be used to compute invariants.

Chapter 8

Predicate/Event-Nets

8.1 An Introductory Example

We consider an example which is well known as "The Dining Philosophers Problem". To start with, we represent it as a C/E-system.

Three philosophers are sitting around a round table. Each philosopher has a plate in front of him. Between any two neighbouring plates lies a fork (Fig. 91). Whenever a philosopher eats he uses both forks, the one to the right and the other to the left of his plate. When a philosopher has finished eating he replaces both his forks on the table and starts thinking. Figure 92 shows this as a C/E-system using the following conditions: d_i (philosopher p_i is thinking), e_i (philosopher p_i is eating) and g_i (the i-th fork is not being used). In the case represented, p_1 is eating, the other two philosophers are thinking and only fork 3 is not being used. Now, the thinking philosophers have to wait until p_1 puts the forks back (u_1) and starts to think. Then a conflict over fork 3 arises and either p_2 or p_3 may start to eat, or p_1 starts eating again.

The three conditions d_i (philosopher p_i is thinking) ($i = 1, 2, 3$) are now combined into one predicate d ("thinking philosophers"). For each case c of the system, it must now be specified for which philosophers the predicate d is true. We now represent the predicate d as an S-element and the philosophers p_i as tokens and mark d with those philosophers for which d is true. Figure 93 illustrates this step.

Analogously, we construct the predicate e ("eating philosophers") and the predicate g ("available forks"). The set of objects for which some predicate is

Fig. 91. The dining philosophers

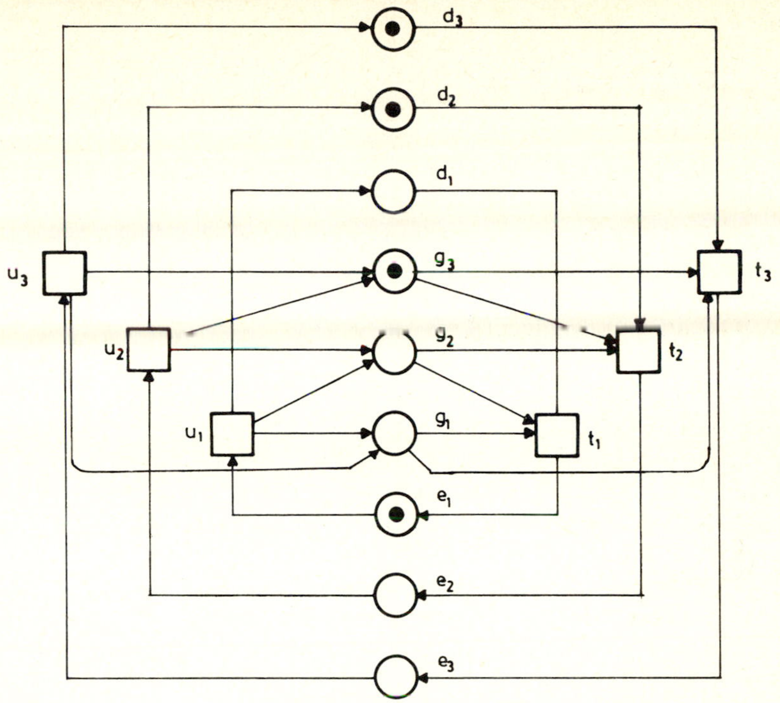

Fig. 92. A C/E-system of the philosophers

true may be modified by events. Such events are again represented as T-elements connecting the predicates. The arcs are labelled to indicate which objects are affected by an event. In this way we obtain the representation in Fig. 94, equivalent to the system shown in Fig. 92.

In Fig. 94, the events t_1, t_2 and t_3 have equal pre- and postsets; they only differ with respect to the affected objects. These three events can be represented by one single T-element as shown in Fig. 95; the affected sets of objects are indicated by arc inscriptions consisting of variables and functions. The functions l and r associate with each philosopher his left and his right fork, respectively. It is possible to derive the concrete events t_i $(1 \leq i \leq 3)$ from the event schema t by substituting for the variable x the respective philosopher p_i. Correspondingly, the T-element u in Fig. 95 is a unification of the events u_1, u_2 and u_3 of Fig. 94.

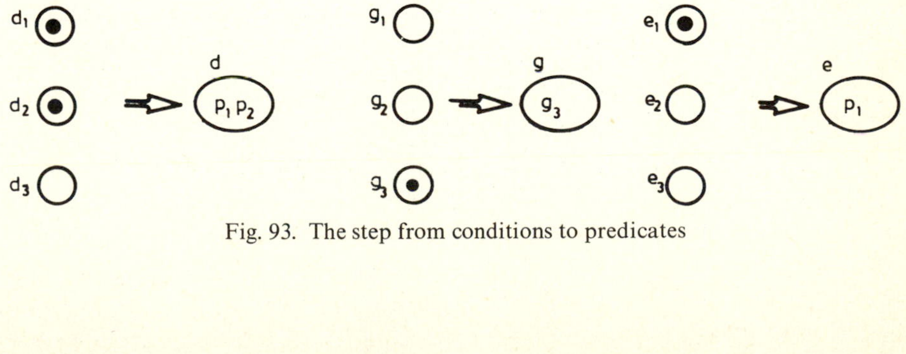

Fig. 93. The step from conditions to predicates

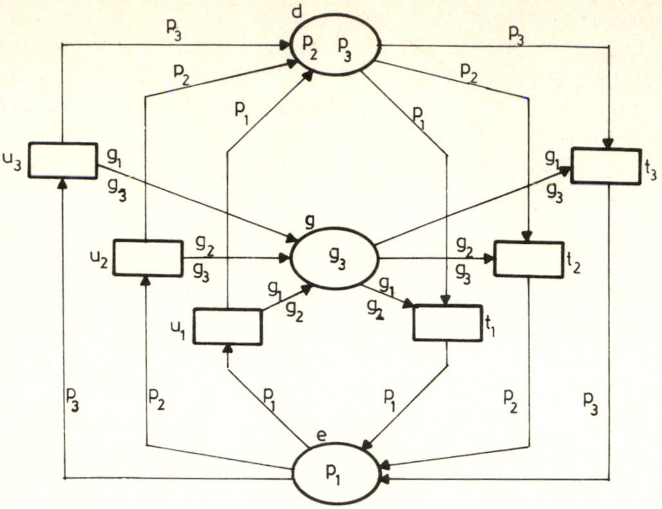

Fig. 94. The system of philosophers using predicates

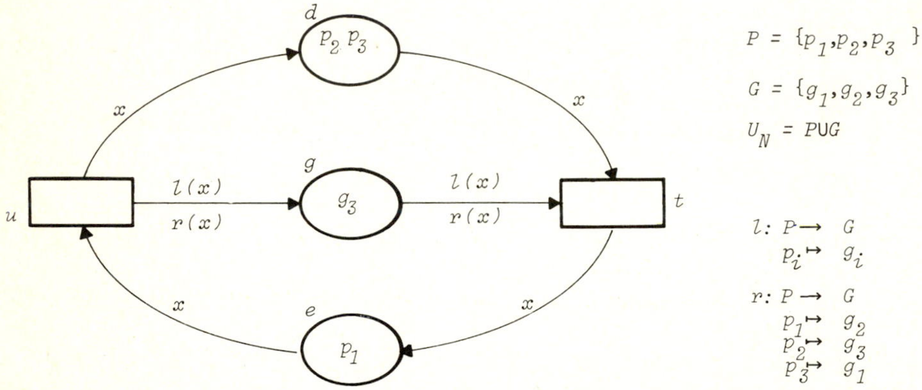

Fig. 95. The system of philosophers using predicates and event schemas

Using the representation shown in Fig. 95 we are able to model a meal of arbitrarily many philosophers: in the initial case let d be marked by $\{p_1, \ldots, p_n\}$, and g by $\{g_1, \ldots, g_n\}$. Now the two functions l and r are defined as $l(p_i) = g_i$ $(i = 1, \ldots, n)$, $r(p_i) = g_{i+1}$ $(i = 1, \ldots, n-1)$ and $r(p_n) = g_1$.

8.2 Predicate/Event-Nets

Now we are going to precisely formulate the concepts introduced informally in 8.1. We start with algebras and define terms over algebras which we shall use as arc inscriptions in the definition of predicate/event-nets.

(a) Definition. Let D be an arbitrary set.

(i) For $n \in \mathbb{N}$ and $M \subseteq D^n$, $f\colon M \to D$ is called a *partial operation* on D. Let Φ be a set of partial operations on D. Then $\underline{D} = (D; \Phi)$ is called an *algebra.* In particular, Φ may contain functions $d\colon D^0 \to D$, which may be identified with elements of D.

(ii) Let X be a set of variables. The set $\mathscr{T}(\underline{D}, X)$ of *terms over D and X* is the smallest set of expressions such that

(a) $X \subseteq \mathscr{T}(\underline{D}, X)$,

(b) if $t_1, \ldots, t_n \in \mathscr{T}(\underline{D}, X)$ and $f\colon D^n \to D \in \Phi$ then $f(t_1, \ldots, t_n) \in \mathscr{T}(\underline{D}, X)$. In particular, an element of D which belongs to Φ as a function $D^0 \to D$ is a term.

(iii) A mapping $\beta\colon X \to D$ is called a *valuation* of X. It induces, canonically, a mapping $\beta\colon \mathscr{T}(\underline{D}, X) \to D$ by $\beta(f(t_1, \ldots, t_n)) = f(\beta(t_1), \ldots, \beta(t_n))$. Finally, we expand β for sets of terms $\mathscr{M} \subseteq \mathscr{T}(\underline{D}, X)$ by $\beta(\mathscr{M}) = \{\beta(t) \mid t \in \mathscr{M}\}$.

Using these notions, we are now able to define the class of nets we discussed informally in the previous section and for which Fig. 95 shows an example.

(b) Definition. $N = (P, E; F, \underline{D}, \lambda, c)$ is called a *predicate/event-net* (*P/E-net*) iff

(i) $(P, E; , F)$ is a net without isolated elements, the elements of P and E are called *predicates* and *events,* respectively,

(ii) $\underline{D}$ is an algebra,

(iii) $\lambda\colon F \to \mathscr{P}(\mathscr{T}(\underline{D}, X)) \setminus \{\emptyset\}$ is a mapping, ($\mathscr{P}$ denotes powerset),

(iv) $c\colon P \to \mathscr{P}(D)$ is the *initial case* of N.

We denote the six components of a P/E-net N by P_N, E_N, F_N, $\underline{D}_N$, λ_N, c_N. In the following we assume the set of variables X and write $\mathscr{T}(N)$ for $\mathscr{T}(\underline{D}_N, X)$, and $\bar{f}$ for $\lambda_N(f)$ $(f \in F_N)$. In Fig. 95, the sets of terms, $\bar{f}$, are written without brackets.

To decide whether an event e of a P/E-net is enabled, one has to consider valuations β and to apply them to the arc inscriptions around e. For arcs (p, e) the set $\beta(\overline{p, e})$ must be contained in the marking of p, for arcs (e, p) no element of $\beta(\overline{e, p})$ may already be contained in the marking of p. When e occurs, the elements of $\beta(\overline{p, e})$ are removed from the predicates $p \in {}^{\bullet}e$, and the elements of $\beta(\overline{e, p})$ are added to the predicates $p \in e^{\bullet}$. Figure 96 shows an example.

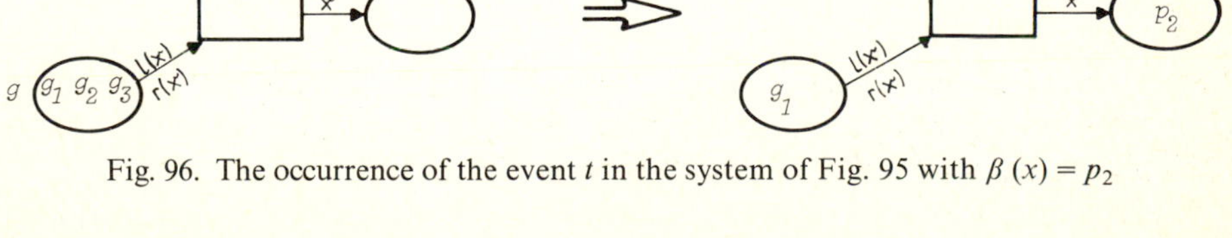

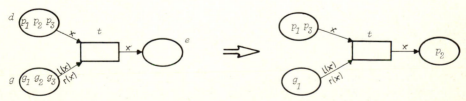

Fig. 96. The occurrence of the event t in the system of Fig. 95 with $\beta(x) = p_2$

(c) Definition. Let N be a P/E-net.

(i) A mapping $c: P_N \to \mathscr{P}(D_N)$ is called a *case* (by analogy with C/E-systems).

(ii) Let $e \in E_N$ and let β be a valuation such that for all $f \in F_N \cap (P_N \times \{e\} \cup \{e\} \times P_N)$: if $t_1, t_2 \in \lambda(f)$ and $t_1 \neq t_2$ then $\beta(t_1) \neq \beta(t_2)$. For a given case c, e is called *c-enabled with* β iff $\forall p \in {}^\bullet e$: $\beta(\overline{p,e}) \subseteq c(p)$ and $\forall p \in e^\bullet$: $\beta(\overline{e,p}) \cap c(p) = \emptyset$.

(iii) An event e which is c-enabled with β yields a *follower case* c' of c under β by

$$c'(p) = \begin{cases} c(p) \setminus \beta(\overline{p,e}) & \text{iff } \; p \in {}^\bullet e \setminus e^\bullet, \\ c(p) \cup \beta(\overline{e,p}) & \text{iff } \; p \in e^\bullet \setminus {}^\bullet e, \\ c(p) \setminus \beta(\overline{p,e}) \cup \beta(\overline{e,p}) & \text{iff } \; p \in {}^\bullet e \cap e^\bullet, \\ c(p) & \text{otherwise.} \end{cases}$$

We say, *e transforms the case c to c' under* β, and we write $c\,[e\rangle_\beta\, c'$. Let $[c_N\rangle$ be the smallest set which contains c_N and which is closed with respect to event occurrences.

To represent a case c graphically, the elements $c(p)$ are written into the circle for p.

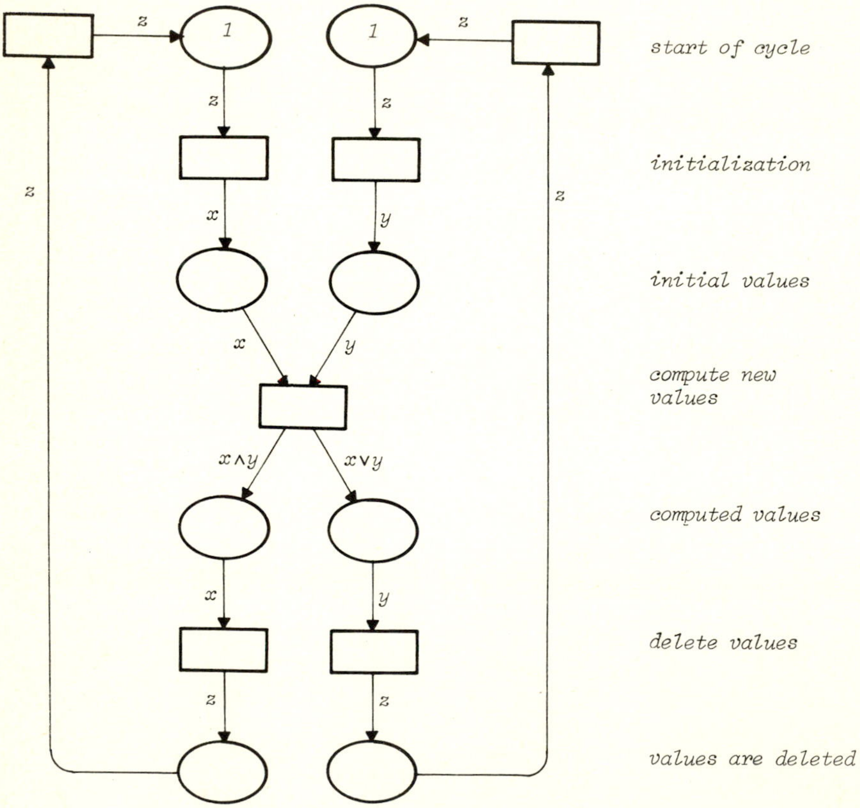

Fig. 97. The System of Fig. 11, represented as a P/E-net with $\underline{D} = (\{0,1\}, \{\vee, \wedge\})$

Figure 97 shows a P/E-net representing the same system as the C/E-system of Fig. 11. The algebra of this P/E-net is the boolean algebra with the carrier $\{0, 1\}$ and the logical operations $\wedge$ and $\vee$.

In most cases, the carrier D of the algebra $\underline{D}_N$ will naturally be the disjunct union of several sets D_i, where each predicate p will only be true for elements of one of these sets D_i, for all reachable cases. For the system represented in Fig. 95 we find that the set of forks belongs, in this way, to the predicate g and the set of philosophers belongs to d and e.

8.3 An Organization Scheme for Distributed Databases

We assume a situation where geographically distributed sites access a common database, in which reading operations occur much more often than writing operations. To minimize the costs of data transmission, it is convenient in this case to have one copy of the data base at each site and to organize an updating mechanism which handles writing operations correctly.

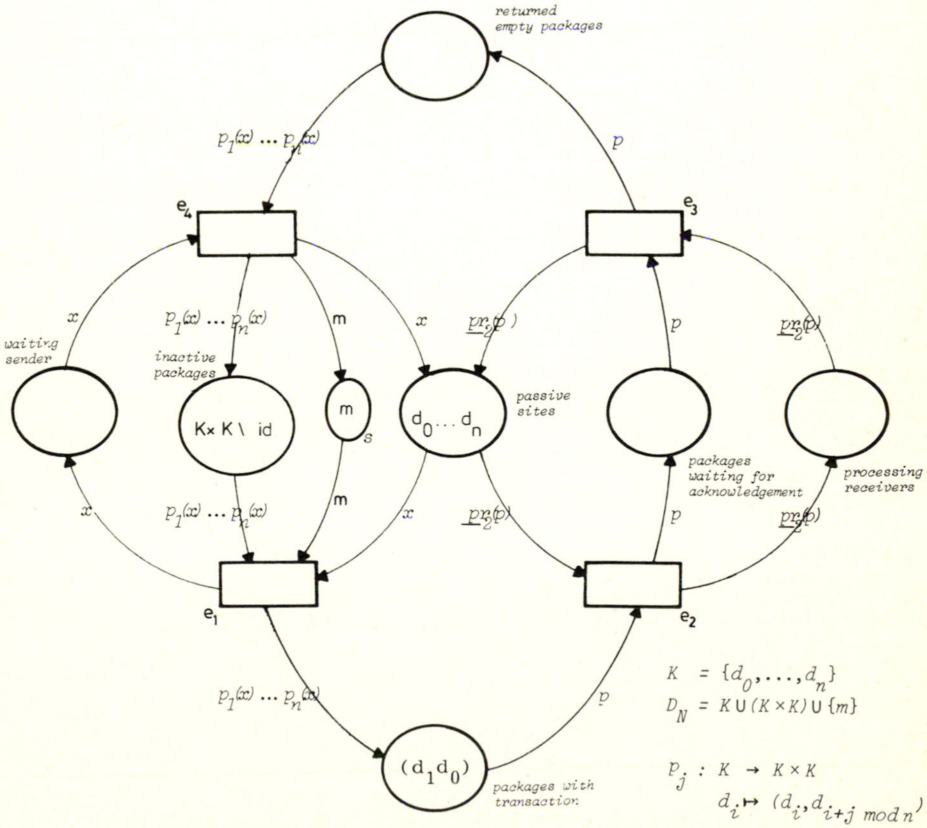

Fig. 98. An organization scheme for updating a distributed database

All updates of the database have to be carried out in the same way in all copies of the database. An update is invoked by a particular site, called the sender, which sends a message to all other sites. Each receiver of such a message updates its copy of the database and sends an acknowledgement back to the sender. The update is successfully completed when the sender has received acknowledgements from all other sites. Since all sites act according to the same scheme we are able to model them as tokens in one single net (Fig. 98).

The message interchange is realized by packages which contain the update message and which are labelled with the identification of the sender and the receiver. Since we are only interested in the organization of the updates and not in the contents of the update messages, we represent each package by a pair consisting of sender and receiver identifications.

Let $K = \{d_0, \ldots, d_n\}$ be the set of involved sites. As long as no messages are being interchanged, the predicate "idle component" is true for all sites d_i and the predicate "inactive package" is true for all packages (d_i, d_j) (see Fig. 98). A site $d_i \in K$ invokes an update procedure by occurrence of the event e_1 with $\beta(x) = d_i$. Then all packages $p_1(d_i), \ldots, p_n(d_i)$ with $p_j(d_i) = (d_i, d_{(i+j) \bmod n})$ are

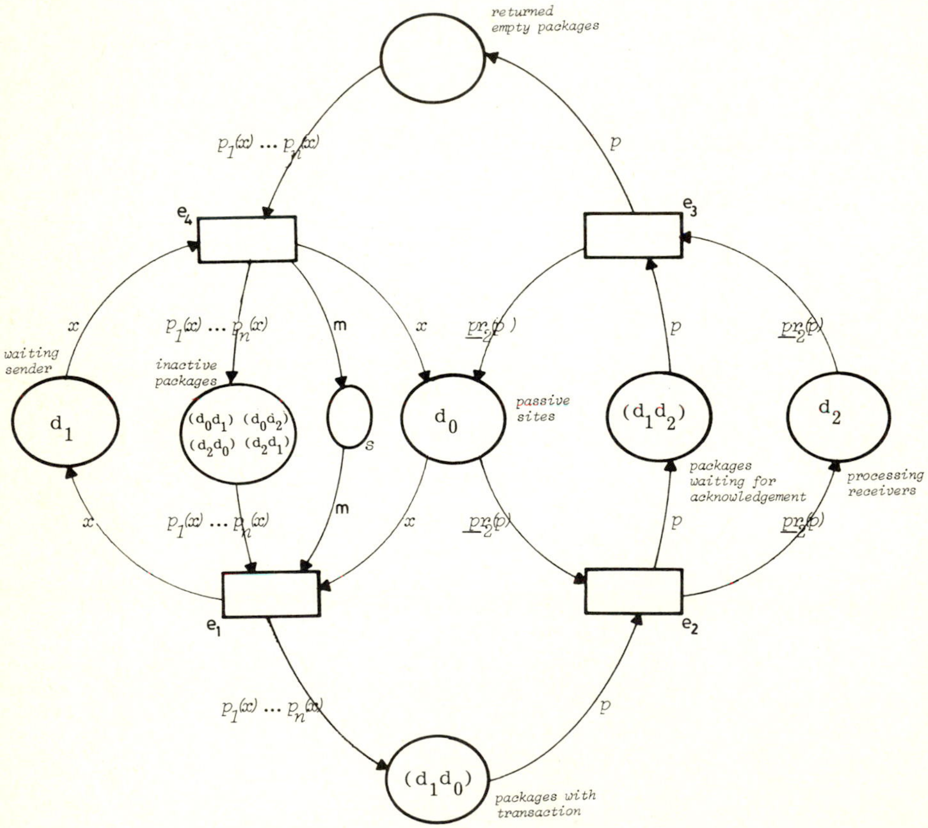

Fig. 99. A case of the system of Fig. 98 with $n = 2$

initialized. d_i is now waiting for the acknowledgements. The event e_2 causes the receivers of the packages to accept and to process the message, independently of each other. When the processing is finished each receiver gives the "empty" package back to the sender by e_3. After all acknowledgements for processed updates have arrived in the form of empty packages, the sender returns to its idle state with e_4. Thereby the packages become inactive. Additionally a token m is put onto s which enables a new cycle.

In Fig. 99 a reachable case is shown for $n = 2$.

8.4 Facts in P/E-Nets

The set of objects for which some predicate of a P/E-net is true changes by event occurrences. Nevertheless there may be relations between predicates which hold for all cases. By analogy with C/E-systems, such relations may be expressed as logical formulae and may be represented as T-elements which are never enabled. In this section we shall derive, as in 4.5, a dependency between the validity of logical formulae and the possibility of events to be enabled.

First, we define those logical formulae which we need to build facts. In terms of the predicate calculus, we have first order formulae in prenex normal form without existential quantifiers. The universal quantifiers may then be omitted.

(a) Definition. Let N be a P/E-net.

(i) The set $\mathscr{A}_N$ of *(logical) formulae over* N is the smallest set such that

(a) if $t \in \mathscr{T}(N)$ and $p \in P_N$ then $p(t) \in \mathscr{A}_N$,

(b) if $a_1, a_2 \in \mathscr{A}_N$ then $(a_1 \wedge a_2) \in \mathscr{A}_N$, $(a_1 \vee a_2) \in \mathscr{A}_N$, $(a_1 \rightarrow a_2) \in \mathscr{A}_N$, $(\neg a_1) \in \mathscr{A}_N$.

As in 4.5, unnecessary brackets will be omitted.

(ii) Each case c of N induces, for each formula $a \in \mathscr{A}_N$ and each valuation β, a value $c_\beta(a) \in \{0, 1\}$, defined by

$$\begin{aligned} c_\beta \colon \mathscr{A}_N &\rightarrow \{0,1\} \\ p(t) &\mapsto \begin{cases} 1 \text{ iff } \beta(t) \in c(p) & (1 \equiv \text{true}), \\ 0 \text{ iff } \beta(t) \notin c(p) & (0 \equiv \text{false}), \end{cases} \\ a_1 \wedge a_2 &\mapsto \min\{c_\beta(a_1), c_\beta(a_2)\}, \\ a_1 \vee a_2 &\mapsto \max\{c_\beta(a_1), c_\beta(a_2)\}, \\ a_1 \rightarrow a_2 &\mapsto c_\beta(\neg a_1 \vee a_2), \\ \neg a &\mapsto 1 - c_\beta(a). \end{aligned}$$

(iii) For each case c of N, let the function $\hat{c}$ be defined as $\hat{c} \colon \mathscr{A}_N \rightarrow \{0, 1\}$,

$$\text{where } \hat{c}(a) = \begin{cases} 1 & \text{iff, for all valuations } \beta,\ c_\beta(a) = 1, \\ 0 & \text{otherwise.} \end{cases}$$

(iv) Two formulae $a_1, a_2 \in \mathscr{A}_N$ are called *equivalent* (we write $a_1 \equiv a_2$) iff, for each case c of N, $\hat{c}(a_1) = \hat{c}(a_2)$.

By analogy with Chap. 4.5, we construct for each event e of a P/E-net N a formula $\alpha(e)$ such that $\alpha(e)$ is true in all cases in which e is not enabled under any valuation β. This will be used in the fact calculus.

(b) Definition. Let N be a finite P/E-net, let $p \in P_N$ and let $e \in E_N$.

(i) For $(p,e) \in F_N$ and $\overline{(p,e)} = \{t_1, \dots, t_k\}$, $\alpha(p,e)$ denotes the formula $p(t_1) \wedge \dots \wedge p(t_k)$.

(ii) For $(e,p) \in F_N$ and $\overline{(e,p)} = \{t_1, \dots, t_g\}$, $\alpha(e,p)$ denotes the formula $p(t_1) \vee \dots \vee p(t_g)$.

(iii) Let ${}^\cdot e = \{p_1, \dots, p_n\}$ and $e^\cdot = \{p_{n+1}, \dots, p_m\}$. Then $\alpha(e)$ is the formula $(\alpha(p_1,e) \wedge \dots \wedge \alpha(p_n,e)) \to (\alpha(e,p_{n+1}) \vee \dots \vee \alpha(e,p_m))$.

(iv) If ${}^\cdot e = \emptyset$ and $e^\cdot = \{p_1, \dots, p_m\}$ then $\alpha(e)$ is the formula $\alpha(e,p_1) \vee \dots \vee \alpha(e,p_m)$.

(v) If ${}^\cdot e = \{p_1, \dots, p_n\}$ and $e^\cdot = \emptyset$ then $\alpha(e)$ is the formula $\neg(\alpha(p_1,e) \wedge \dots \wedge \alpha(p_n,e))$.

In Fig. 95, we have: $\alpha(u) \equiv e(x) \to d(x) \vee g(l(x)) \vee g(r(x))$,
$\alpha(t) \equiv d(x) \wedge g(l(x)) \wedge g(r(x)) \to e(x)$.

(c) Theorem. *Let N be a finite P/E-net and let $e \in E_N$. Then, for each case $c \in [c_N\rangle$: $\hat{c}(\alpha(e)) = 1$ iff e is not c-enabled with any valuation β.*

Proof. $\hat{c}(\alpha(e)) = 1 \Leftrightarrow \forall\beta: c_\beta(\alpha(e)) = 1$

$\Leftrightarrow \forall\beta: (\exists p \in {}^\cdot e$ with $c_\beta(\alpha(p,e)) = 0 \vee \exists p \in e^\cdot$ with $c_\beta(\alpha(e,p)) = 1)$

$\Leftrightarrow \forall\beta: (\exists p \in {}^\cdot e$ and $\exists t \in \overline{(p,e)}$ with $c_\beta(p(t)) = 0$
$\vee\, \exists p \in e^\cdot$ and $\exists t \in \overline{(e,p)}$ with $c_\beta(p(t)) = 1)$

$\Leftrightarrow \forall\beta: (\exists p \in {}^\cdot e\ \exists t \in \overline{(p,e)}$ with $\beta(t) \notin c(p)$
$\vee\, \exists p \in e^\cdot\ \exists t \in \overline{(e,p)}$ with $\beta(t) \in c(p)$

$\Leftrightarrow \forall\beta: (\exists p \in {}^\cdot e$ with $\beta(\overline{(p,e)}) \nsubseteq c(p) \vee \exists p \in e^\cdot$ with $\beta(\overline{(e,p)}) \cap c(p) \neq \emptyset)$

$\Leftrightarrow$ e is not c-enabled with any valuation β. □

By analogy with 4.5, we saw in the previous section that T-elements which never become enabled represent formulae which are valid for all cases. Now we shall show that each valid formula may be represented by such T-elements.

(d) Definition. Let N be a P/E-net.

(i) A formula $\alpha \in \mathscr{A}_N$ is called *valid in N* iff, for all cases $c \in [c_N\rangle$, $\hat{c}(\alpha) = 1$.

(ii) For $P_1, P_2 \subseteq P_N$ and $P_1 \cup P_2 \neq \emptyset$, let $t = (P_1, P_2)$ be a new T-element with ${}^\cdot t = P_1$ and $t^\cdot = P_2$. For each new arc $f \in (P_1 \times \{t\}) \cup (\{t\} \times P_2)$, let a set of terms $\lambda(f) \subseteq \mathscr{T}(N)$ be given by a mapping λ. (t, λ) is called a *fact of N* iff t is never enabled for any case $c \in [c_N\rangle$ and any valuation β.

(iii) Corresponding to Definition 8.2 (b), we also use with respect to a fact t the notations $\overline{(p,t)}$ and $\overline{(t,p)}$ for $\lambda(p,t)$ and $\lambda(t,p)$, respectively. The formula $\alpha(t)$ is defined as $\alpha(e)$ for events e.

In the graphical representation of P/E-nets, a fact is drawn as □, as for C/E-systems; the associated arcs are appropriately inscribed.

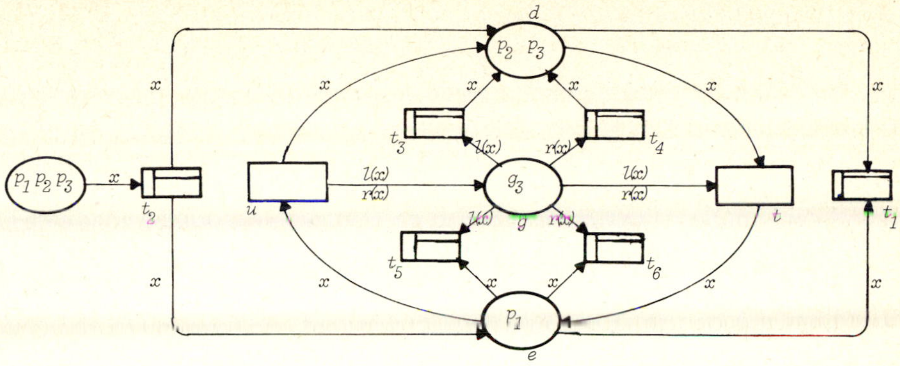

Fig. 100. Some facts in the system of Fig. 95

Figure 100 shows some facts in the system of the dining philosophers. Their meaning may be explained as follows:

t_1: $\neg(d(x) \wedge e(x))$: A thinking philosopher is not eating and an eating philosopher is not thinking.
t_2: $p(x) \rightarrow d(x) \vee e(x)$: Each philosopher is either eating or thinking. There is no other activity represented in this system.
t_3: $g(l(x)) \rightarrow d(x)$: Whenever the left fork of some philosopher is not being used then he is thinking.
t_4: As t_3, with respect to the right fork.
t_5: $\neg(e(x) \wedge g(l(x)))$: whenever a philosopher is eating then his left fork is not available, and whenever his left fork is available he is not eating.
t_6: As t_5, with respect to the right fork.

(e) Theorem. *Let N be a P/E-net and let $a \in \mathcal{A}_N$. a is valid in N iff there exist facts $t_1, \ldots, t_n$ such that a is logically equivalent to $a(t_1) \wedge \ldots \wedge a(t_n)$.*

Proof. The if-part follows immediately using Theorem 8.4 (c). Conversely, a can be transformed into a logical equivalent formula $a' = a_1 \wedge \ldots \wedge a_k$ in conjunctive normal form. Each a_g $(1 \le g \le k)$ is a term of the form $\neg q_1(t_1) \vee \ldots \vee \neg q_n(t_n) \vee q_{n+1}(t_{n+1}) \vee \ldots \vee q_m(t_m)$ with $q_1, \ldots, q_m \in P_N$ and $t_1, \ldots, t_m \in \mathcal{T}(N)$. For each $p \in P_N$, let $\mathcal{T}_p = \{t_i \,|\, 1 \le i \le n \wedge q_i = p\}$ and $\bar{\mathcal{T}}_p = \{t_j \,|\, n+1 \le j \le m \wedge q_j = p\}$. Now let t_g be a new element with ${}^{\bullet}t_g = \{p \,|\, \mathcal{T}_p \neq \emptyset\}$ and $t_g^{\bullet} = \{p \,|\, \bar{\mathcal{T}}_p \neq \emptyset\}$, and let $\overline{(p, t_g)} = \mathcal{T}_p$ and $\overline{(t_g, p)} = \bar{\mathcal{T}}_p$. Clearly, $a_g = a(t_g)$ $(g = 1, \ldots, k)$. Hence each t_g is a fact and a is logically equivalent to $a(t_1) \wedge \ldots \wedge a(t_g)$. □

As for *C/E*-systems, we again have the problem of how to verify facts. We shall see that, for *P/E*-nets also, the concept of invariants is helpful for this. To deal with invariants, we shall introduce a slightly different net model, called "relation nets". *P/E*-nets are transformable into relation nets, using the normal form which is introduced in the next section.

8.5 A Normal Form for *P/E*-Nets

In the normal form we are going to construct, we shall reduce the number of variables in the environment of events. Instead of variables $x_1, \ldots, x_n$ which are valuated by single elements of D_N we use one variable x which is now valuated by objects of $(D_N)^n$. The variables x_i are then simulated by projections. So only the arc inscriptions have to be changed and other valuations of the variables have to be used for event occurrences.

(a) Definition. Let N be a finite P/E-net and let $X = \{x_1, \ldots, x_n\}$ be the variables occuring in terms of N.

(i) With $D_N = (D, \Phi)$, let $\hat{\underline{D}} := (D \cup D^n, \Phi \cup \{\underline{pr}_i \mid \leq i \leq n\})$. We associate with each term $t \in \mathcal{T}(N)$ a term $\hat{t} \in \mathcal{T}(\hat{\underline{D}}, \{x\})$ in the following way:
$\hat{t} = \underline{pr}_i(x)$ iff $t = x_i$ $(1 \leq i \leq n)$,
$\hat{t} = \underline{f}(\hat{t}_1, \ldots, \hat{t}_n)$ iff $t = \underline{f}(t_1, \ldots, t_n)$.

(ii) Let $\hat{N} = (P_N, E_N; F_N, \hat{\underline{D}}, \lambda, c_N)$, where $\lambda(f) = \{\hat{t} \mid t \in \lambda_N(f)\}$.

(b) Definition. A P/E-net is called in *normal form* iff $|X| = 1$.

(c) Corollary. *Let N be a finite P/E-net. Then $\hat{N}$ is in normal form.*

The net shown in Fig. 95 is in normal form. Figure 101 shows a scheme for the construction of the normal form.

(d) Definition. Two P/E-nets N and N' are called *equivalent* iff $D_N = D_{N'} \wedge E_N = E_{N'} \wedge F_N = F_{N'} \wedge (\forall c_1, c_2 \in [c_N\rangle, \ \forall e \in E_N$: there exists a valuation β with $c_1\,[e\rangle_\beta\, c_2$ in N iff there exists a valuation β' with $c_1\,[e\rangle_{\beta'}\, c_2$ in N').

(e) Lemma. *Each P/E-net N is equivalent to its normal form $\hat{N}$.*

Proof. Let $X = \{x_1, \ldots, x_n\}$ be the set of variables of N and let x be the variable of $\hat{N}$. We associate with the valuation β induced by $\beta: X \to D$ a valuation β' by $\beta'(x) = (\beta(x_1), \ldots, \beta(x_n))$. Conversely, if $\beta': \{x\} \to D$ is given, then let β be defined by $\beta(x_i) = \underline{pr}_i(\beta'(x))$. □

The library system is represented in Fig. 18 as a P/E-net with events labelled by additional inscriptions. Such inscriptions may be considered as ad-

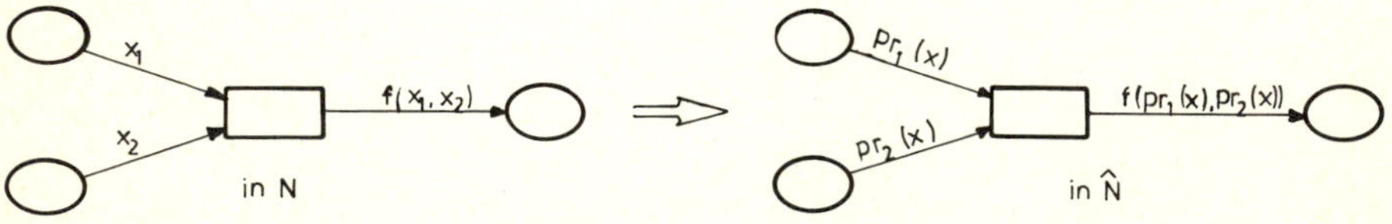

Fig. 101. Construction of the normal form of a *P/E*-net

ditional predicates which must be fulfilled before an event occurs and which are not changed by the occurrence of this event. In the calculus of *P/E*-nets they have to be represented as additional *S*-elements.

For system design, it is of course recommended that more than one variable is used and that events are labelled by conditions. Such conditions have only to be substituted if system properties are represented as facts. Also the restriction to one variable in the environment of events is necessary only if invariants are to be calculated.

Exercises for Chapter 8

1. Represent the four season system (Fig. 1) as a *P/E*-net with a minimal number of predicates and events.

2. In Fig. 98 represent the following facts:
   a) Whenever a package is waiting for acknowledgement, its corresponding receiver is processing.
   b) Whenever an empty package is to be returned, its sender is waiting.

*3. Supplement the system of dining philosophers (Fig. 95) with a fair schedule such that each philosopher who wants to eat, will eventually be able to eat.

Chapter 9

Relation Nets

After introducing P/E-nets, we now present a further net model using individuals as tokens. This new model, in particular, supports a calculus of invariants.

In Chap. 6 we introduced the idea of invariants for P/T-nets. Now we generalize the notion of markings of P/T-nets to individual tokens in the same way as we generalized the notion of cases of C/E-systems, when defining P/E-nets. A marking will now indicate, for each place, not only the number but also the sorts of its tokens. Thus a marking $M(s)$ of some place s is a mapping $M(s)\colon D \to \mathbb{N}$ giving for each sort $d \in D$ the number of tokens of this sort d on s. Whenever a transition fires, the distribution of the typed tokens over the places is changed.

We recall, from Chap. 6, some prerequisites for the construction of S-invariants. For expressions of the form $\underline{N}' \cdot x = 0$ or $i \cdot M = i \cdot M_N$ to be sensible it must be possible to multiply matrix entries with each other and with markings, the results of these operations have to be summed. With respect to addition, a neutral element "0" is required and the multiplication must be distributive over the addition.

As the arc inscriptions are used as matrix entries, these inscriptions and the whole net model must be chosen in such a way that such operations are possible. As the arcs will be labelled using relations, the resulting nets will be called *relation nets.*

We shall show in which way P/E-nets may be considered as special relation nets. Using a matrix representation, a calculus for S-invariants is obtained. This may be used to verify facts.

9.1 Introductory Examples

We start with the illustration of the main idea underlying the concept of relation nets, by considering a special case. We show how to represent P/E-nets as relation nets. Every P/E-net may be transformed into a relation net in the following way: Each arc inscription $\bar{f}$ of a P/E-net in normal form yields, for each valuation β of the variable x, the set $\beta(\bar{f}) \subseteq D$. Hence we may consider the meaning of $\bar{f}$ as a set of tuples $(\beta(x), y)$ with $y \in \beta(\bar{f})$, i.e. $\bar{f}$ denotes the relation $\{(a, b) \mid \exists$ valuation β with $a = \beta(x)$ and $b \in \beta(\bar{f})\} \subseteq D \times D\}$. A transition t fires with respect to some parameter d by removing, from each place

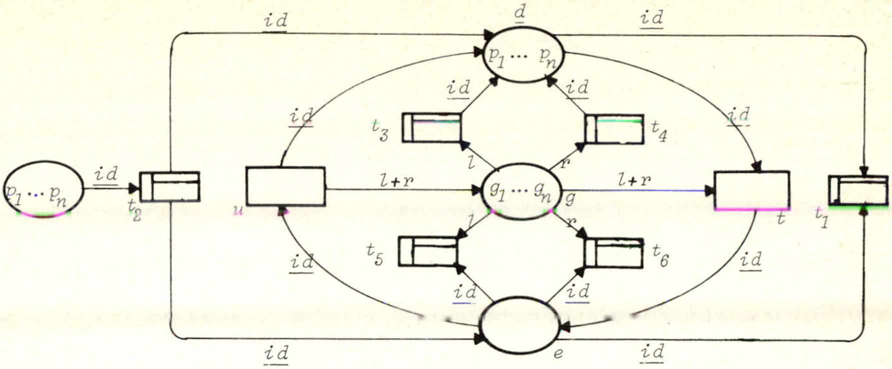

Fig. 102. The dining philosophers represented as a relation net (cf. Fig. 100)

$s \in {}^{\bullet}t$, the elements $(\overline{s,t})\,[d]$ and by adding to each place $s \in t^{\bullet}$, the elements $(\overline{t,s})\,[d]$ (see A6 (iv)). Figure 102 shows a relation net with the same meaning as the net shown in Fig. 100. Thereby the graph of a function is considered as a relation (id denotes the identity relation).

We see that, when constructing a relation net N' from a P/E-net N, markings $M(s) \subseteq D$ are represented by their characteristic mapping $M(s)$: $D \to \{0, 1\}$. Each arc inscription $\bar{f} \subseteq \mathcal{T}(\underline{D}, \{x\})$ of N is transformed into a relation $\bar{f} \subseteq D \times D$ which again may be considered as a characteristic mapping $\bar{f}: D \times D \to \{0, 1\}$. In the general case, we shall have markings of the form $M(s): D \to \mathbb{N}$ and arc inscriptions of the form $\bar{f}: D \times D \to \mathbb{N}$ in relation nets. A transition t fires with respect to some parameter a by removing, from each

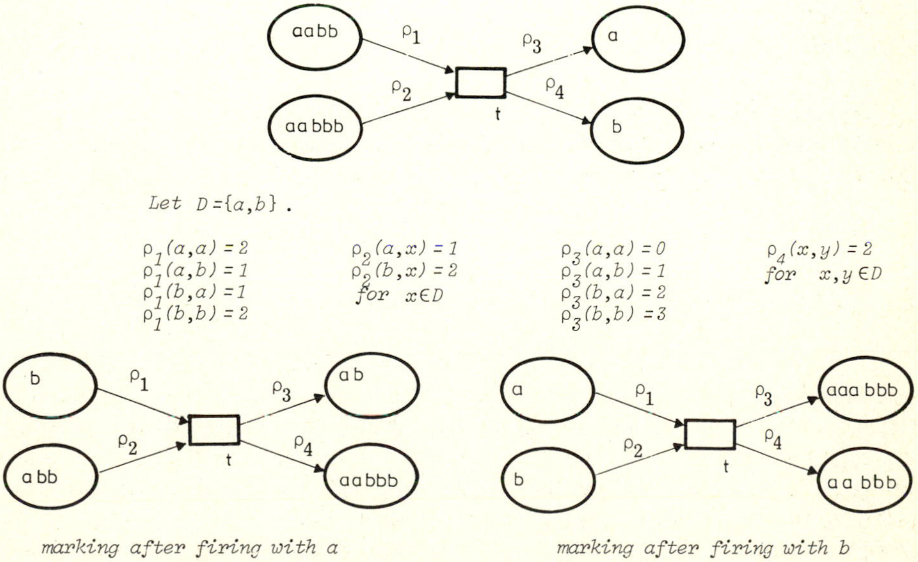

Fig. 103. The firing of a transition, t, of a relation net

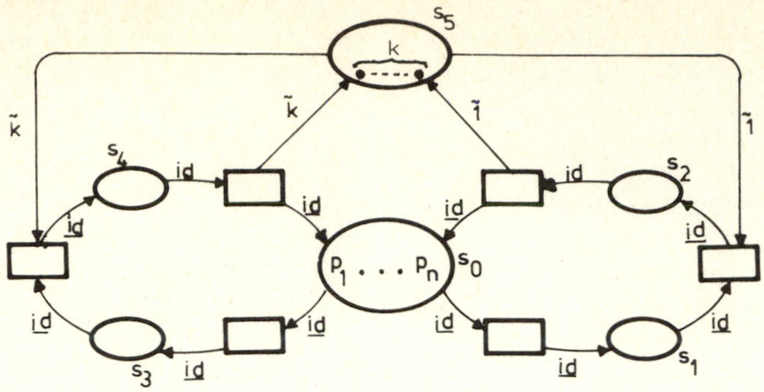

Fig. 104. The system of reader and writer processes of Fig. 66, identifying single processes

place $s \in {}^{\bullet}t$, $(\overline{s,t})\,(a, d)$ elements of each sort $d \in D$ and by adding, to each place $s \in t^{\bullet}$, $(\overline{t,s})\,(a, d)$ elements of each sort $d \in D$.

It is convenient to use relation nets if several individuals of some sort do not have to be distinguished. One should not be forced to distinguish individuals if one doesn't wish to. This would lead to overspecification. The system of reader and writer processes shown in Fig. 66 is an example of this. There it might be convenient to distinguish the processes but it is certainly not necessary to distinguish the k control tokens. Figure 104 shows a representation as a relation net where this is realized. Mappings of the form $A: D \to \mathbb{Z}$ and $\varrho: D \times D \to \mathbb{Z}$ will, in the following, be called *multisets* and *multirelations*, respectively. These names reflect their nature as generalizations of characteristic mappings of sets and relations, respectively.

9.2 Relation Nets

In the previous section, we gave an introduction to the use of multisets and multirelations in relation nets. A multiset M defines for each element, d, of some set of sorts, D, how often d is contained in M. Thereby we allow that some element d may also be contained in M "negatively often". It is therefore possible to calculate with multisets as with integers. In particular, they can be added, subtracted and multiplied with integers by performing the corresponding operations for each sort separately. Multirelations are multisets over the cartesian product $D \times D$ of a set of sorts D.

(a) Definition. Let D be a set.

(i) A *multiset over* D is a mapping $M: D \to \mathbb{Z}$. Let $\mathcal{M}(D)$ denote the set of all multisets over D.

(ii) A multiset $A \in \mathcal{M}(D)$ is called *positive* iff $\forall d \in D: A(d) \geq 0$. Let $\mathcal{M}_+(D)$ denote the set of all positive multisets over D.

(iii) We define the *addition, product with integers* $z \in \mathbb{Z}$ and the *ordering* $\leq$ for multisets $A, B \in \mathcal{M}(D)$ as

$$(\mathcal{M}_1)\quad A+B\colon\ D \to \mathbb{Z}$$
$$d \mapsto A(d) + B(d),$$

$$(\mathcal{M}_2)\quad z \cdot A\colon\ D \to \mathbb{Z}$$
$$d \mapsto z \cdot A(d),$$

$$(\mathcal{M}_3)\quad A \leq B \quad \Leftrightarrow \forall d \in D \quad A(d) \leq B(d).$$

For the handling of multisets in our calculus the following notations and shorthands are convenient:

(b) Definition. Let D be a set.
(i) For $A, B \in \mathcal{M}(D)$, let $-A = (-1) \cdot A$ and $A - B = A + (-B)$.
(ii) For $z \in \mathbb{Z}$, let the multiset $\underline{z} \in \mathcal{M}(D)$ be given by $\underline{z}(d) = z$. In particular, $\underline{0}$ denotes the *empty multiset*.

Multisets A with images $A(d) \in \{0, 1\}$ for all $d \in D$ are (characteristic mappings of) sets. In this case, the addition $+$ may be interpreted as the disjunct union $\uplus$, and the ordering $\leq$ as the inclusion $\subseteq$. If $B \subseteq A$ we then have $A - B = A \backslash B$.

If $D = \{d_1, \ldots, d_n\}$, we shall write multisets $A \in \mathcal{M}(D)$ also as linear combinations $m_1 d_1 + \ldots + m_n d_n$, where $m_i = A(d_i)$ $(i = 1, \ldots, n)$. It is sufficient to specify those elements d_i for which $m_i \neq 0$. In this sense, each summand $m_i d_i$ denotes a multiset M (by $M(d_i) = m_i$, $M(d) = 0$ for $d \neq d_i$) and it is possible to calculate using this representation according to $\mathcal{M}_1, \mathcal{M}_2$ and $\mathcal{M}_3$ as with integer vectors. When the multiplicity 1 is not explicitly written, each element $d \in D$ is a multiset itself and we write D also for the multiset $\sum_{d \in D} d$.

(c) Definition. Let D be a set.
(i) $\mathcal{R}(D) = \mathcal{M}(D^2)$ denotes the set of all *multirelations over* D.
$\mathcal{R}_+(D) = \mathcal{M}_+(D^2)$ denotes the set of all *positive multirelations over* D.
(ii) For $\varrho \in \mathcal{R}(D)$ and $a \in D$, let $\varrho[a]\colon D \to \mathbb{Z}$, $d \mapsto \varrho(a, d)$.
Hence $\varrho[a]$ is a multiset.
(iii) Let $\underline{\mathrm{id}} \in \mathcal{R}_+(D)$ be given by $\underline{\mathrm{id}}(x, y) = 1$ iff $x = y$, $\underline{\mathrm{id}}(x, y) = 0$ iff $x \neq y$.
Let $O \in \mathcal{R}_+(D)$ be defined as $O(x, y) = 0$ for all $x, y \in D$.

As multirelations are special multisets, it is possible to calculate with them according to the rules of 9.2 (a).

Now we define relation nets as nets with positive multisets as markings and positive multirelations as arc inscriptions.

(d) Definition. A 7-tuple $N = (S, T; F, K, D, \lambda, M)$ is called a *relation net* iff
(i) $(S, T; F)$ is a net, the elements of S and T are called *places* and *transitions*, respectively,
(ii) $K: S \to (D \to \mathbb{N} \cup \{\omega\})$ defines a (possibly unlimited) *capacity* for each place,
(iii) D is a set, and $\lambda: F \to \mathcal{R}_+(D)$ associates with each arc a positive multirelation as an inscription,

(iv) $M: S \to \mathcal{M}_+(D)$ is an *initial marking* respecting the capacities, i.e. $\forall s \in S$: $M(s) \leq K(s)$.

Again, we denote the components of a relation net N by $S_N, T_N, F_N, K_N, D_N, \lambda_N$ and M_N. As for P/E-nets we write $\bar{f}$ for $\lambda(f)$.

(e) Definition. Let N be a relation net.

(i) A mapping $M: S_N \to \mathcal{M}_+(D_N)$ is called a *marking of* N iff $\forall s \in S_N: M(s) \leq K_N(s)$.

(ii) For $d \in D_N$ and a marking M of N, a transition $t \in T_N$ is called *M-enabled with* d iff $\forall s \in {}^{\bullet}t: M(s) \geq \overline{(s, t)}[d]$ and $\forall s \in t^{\bullet}: M(s) \leq K_N(s) - \overline{(t, s)}[d]$ and $\sum_{s \in {}^{\bullet}t} \overline{(s, t)}[d] + \sum_{s \in t^{\bullet}} \overline{(t, s)}[d] > \underline{0}$.

(iii) A transition $t \in T_N$ which is M-enabled with d yields a *follower marking* M' of M by

$$M'(s) = \begin{cases} M(s) - \overline{(s, t)}[d] & \text{iff } s \in {}^{\bullet}t \setminus t^{\bullet}, \\ M(s) + \overline{(t, s)}[d] & \text{iff } s \in t^{\bullet} \setminus {}^{\bullet}t, \\ M(s) - \overline{(s, t)}[d] + \overline{(t, s)}[d] & \text{iff } s \in {}^{\bullet}t \cap t^{\bullet}, \\ M(s), \text{ otherwise.} \end{cases}$$

We say *t fires from M to M'* and we write $M[t\rangle_d M'$.

(iv) Let $[M\rangle$ be the smallest set of markings which contains M and which is closed with respect to transition firings.

In the graphical representation arcs f are labelled by $\bar{f}$. A marking M is represented by drawing, into each place s, $M(s)(d)$ tokens of each sort d.

The nets shown in Fig. 102 and Fig. 104 are relation nets. In most cases, the carrier D_N of a relation net N will naturally be the disjunct union of several sets D_i, such that, for each place s, the reachable markings consist only of elements of one set D_i. In the net of Fig. 104, the set of processes $\{p_1, \ldots, p_n\}$ belongs to $s_1, \ldots, s_4$; the k control tokens belong to s_5.

All the different net models considered until now (C/E-systems, P/T-nets, P/E-nets) are special classes of relation nets. Figure 105 shows how the markings have to be restricted to obtain the corresponding special classes.

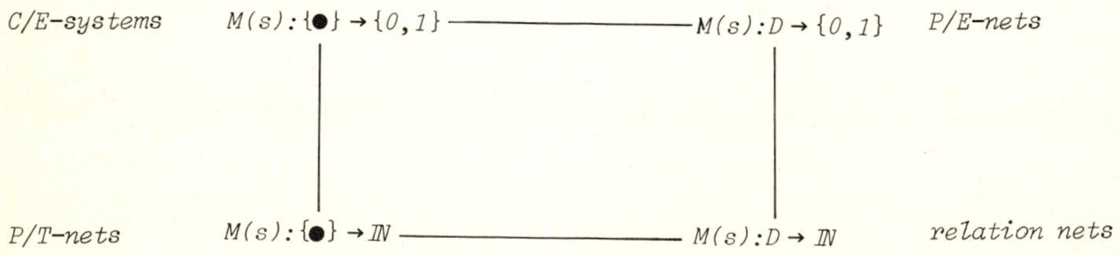

Fig. 105. Relations between different net models

9.3 The Translation of P/E-Nets into Relation Nets

In Chap. 8.5, we derived a normal form for P/E-nets. Now we shall associate with each P/E-net a behaviourly equivalent relation net. The idea of this construction has already been introduced in Sect. 9.1.

(a) Definition.

(i) Let $\underline{D}$ be an algebra. For terms $t \in \mathscr{T}(\underline{D}, \{x\})$, let the multirelation $\varrho(t): D^2 \to \{0,1\}$ be given by $\varrho(t)(a,b) = 1$ iff there exists a valuation β with $a = \beta(x)$ and $b = \beta(t)$.

(ii) Let N be a P/E-net in normal form with terms over $\{x\}$. Let $\varrho_N: F_N \to \mathscr{R}_+(D_N)$ be defined as $\varrho_N(f) = \sum_{t \in \bar{f}} \varrho(t)$. By Definition 8.2 (b) we have $\varrho_N(f)(a,b) \leq 1$. For arbitrary cases c of N, let $M_c: P_N \to \mathscr{M}_+(D_N)$ be defined as $M_c(p)(d) = 1$ iff $d \in c(p)$, and $M_c(p)(d) = 0$, otherwise.

(b) Theorem. *Let $N = (P, E; F, \underline{D}, \lambda, c)$ be a P/E-net in normal form and let the relation net N' be given by $N' = (P, E; F, K, D, \varrho_N, M_c)$ with $\forall p \in P: K(p) = \underline{1}$. Then, $\forall c, c' \in C_N$ $\forall e \in E$ $\forall$ valuations $\beta: \{x\} \to D: c\,[e\rangle_\beta\, c'$ in N iff $M_c[e\rangle_{\beta(x)}$ $M_{c'}$ in N'.*

Proof. Let $p \in {}^\bullet e \setminus e^\bullet$. $c'(p) = c(p) \setminus \beta(\overline{p,e})$
$\Leftrightarrow (\forall d \in \beta(\overline{p,e}): d \in c(p) \wedge d \notin c'(p)$ and, $\forall d \notin \beta(\overline{p,e}): d \in c(p) \Leftrightarrow d \in c'(d))$
$\Leftrightarrow (\forall d \in D$ with $\varrho_N(p,e)(\beta(x), d) = 1: M_c(p)(d) = 1 \wedge M_{c'}(p)(d) = 0$ and, $\forall d \in D$ with $\varrho_N(p,e)(\beta(x), d) = 0: M_c(p)(d) = M_{c'}(p)(d))$
$\Leftrightarrow M_{c'}(p) = M_c(p) - \varrho_N(p,e)[\beta(x)]$.

Analogously, $\forall p \in e^\bullet \setminus {}^\bullet e: c'(p) = c(p) \cup \beta(\overline{p,e})$ $\Leftrightarrow$ $M_{c'}(p) = M_c(p) + \varrho_N(p,e)[\beta(x)]$. By a similar treatment of the remaining cases the result follows. □

9.4 Calculation with Multirelations

For the calculus of invariants of relation nets, we shall use a matrix representation. Therefore it must be possible to add and to multiply matrix entries, which are arc inscriptions and hence multirelations. The summation of multirelations was previously defined in Chap. 9.2. Now we shall define a product, which turns out to be a generalization of the relation product (see A6 (ii)) to multirelations. In particular, we shall show that addition is distributive for the product. Finally we shall introduce a calculus for vectors consisting of multisets and multirelations. At this point we shall have prepared all the prerequisites for the calculus of invariants.

We start by presenting all operations for multirelations.

In 9.2, we defined the application of a multirelation ϱ to a single element $a \in D$ to be the multiset $\varrho[a]$ with $\varrho[a](d) = \varrho(a, d)$. Applying ϱ to a multiset A yields a multiset $\varrho[A]$ as follows: To determine $\varrho[A](d)$, we consider, for each $e \in D$, the integer $A(e)$ as a factor modifying $\varrho(e, d)$.

So $A(e)\,\varrho(e, d)$ yields the contribution of e to $\varrho[A](d)$. $\varrho[A](d)$ is obtained as the sum of all products of this form.

The composition $\varrho \circ \sigma$ of two multirelations ϱ and σ is again a multirelation. To compute $\varrho \circ \sigma(a, b)$, we consider, for each $e \in D$, the integers $\varrho(a, e)$ and $\sigma(e, b)$. Their product yields the contribution of e to $\varrho \circ \sigma(a, b)$. $\varrho \circ \sigma(a, b)$ is the sum of all products of this form.

(a) Definition. Let D be a set. Let $a \in D$, $A \in \mathcal{M}(D)$, $\varrho, \sigma \in \mathcal{R}(D)$ and $z \in \mathbb{Z}$.

The addition and the product with integers for multirelations are given as the corresponding multiset operations:

$(\mathcal{R}_1)$ $\varrho + \sigma: D^2 \to \mathbb{Z}$, $(a, b) \mapsto \varrho(a, b) + \sigma(a, b)$ $\qquad$ $(\mathcal{R}_2)$ $z \cdot \varrho: D^2 \to \mathbb{Z}$, $(a, b) \mapsto z \cdot (\varrho(a, b))$.

The application of a multirelation to a single element and to a multiset, respectively, are defined as

$(\mathcal{R}_3)$ $\varrho[a]: D \to \mathbb{Z}$, $d \mapsto \varrho(a, d)$ $\qquad$ and $\qquad$ $(\mathcal{R}_4)$ $\varrho[A]: D \to \mathbb{Z}$, $d \mapsto \sum_{e \in D} A(e) \cdot \varrho(e, d)$.

The composition of multirelations is defined as

$(\mathcal{R}_5)$ $\varrho \circ \sigma: D^2 \to \mathbb{Z}$, $(a, b) \mapsto \sum_{e \in D} \varrho(a, e) \cdot \sigma(e, b)$.

As an example, with the multiset A, defined as $A(a) = 2$ and $A(b) = -1$, we find using the relations ϱ_1 and ϱ_3 of Fig. 103:

$$\varrho_1[A](a) \stackrel{\mathcal{R}_4}{=} A(a) \cdot \varrho_1(a, a) + A(b) \cdot \varrho_1(b, a) = 2 \cdot 2 - 1 \cdot 1 = 3,$$
$$\varrho_1[A](b) \stackrel{\mathcal{R}_4}{=} A(a) \cdot \varrho_1(a, b) + A(b) \cdot \varrho_1(b, b) = 2 \cdot 1 - 1 \cdot 2 = 0,$$
$$\varrho_1 \circ \varrho_3(a, b) \stackrel{\mathcal{R}_5}{=} \varrho_1(a, a) \cdot \varrho_3(a, b) + \varrho_1(a, b) \cdot \varrho_3(b, b) = 2 \cdot 1 + 1 \cdot 3 = 5,$$
$$\varrho_1 \circ \varrho_3(b, a) \stackrel{\mathcal{R}_5}{=} \varrho_1(b, a) \cdot \varrho_3(a, a) + \varrho_1(b, b) \cdot \varrho_3(b, a) = 1 \cdot 0 + 2 \cdot 2 = 4.$$

(b) Lemma. *Let D be a set. Let $a \in D$, $A, B \in \mathcal{M}(D)$, $\varrho, \sigma, \tau \in \mathcal{R}(D)$ and $z \in \mathbb{Z}$.*

(i) $A + B = B + A$
(ii) $\varrho + \sigma = \sigma + \varrho$
(iii) $z \cdot (\varrho + \sigma) = (z \cdot \varrho) + (z \cdot \sigma)$
(iv) $\varrho \circ (z \cdot \sigma) = z \cdot (\varrho \circ \sigma)$
(v) $(\varrho + \sigma)[a] = \varrho[a] + \sigma[a]$
(vi) $\varrho[A + B] = \varrho[A] + \varrho[B]$
(vii) $(\varrho \circ \sigma)[a] = \sigma[\varrho[a]]$
(viii) $\varrho \circ (\sigma + \tau) = (\varrho \circ \sigma) + (\varrho \circ \tau)$

Proof. Let $a, b \in D$.

(i) $(A + B)(a) \stackrel{\mathcal{M}_1}{=} A(a) + B(a) = B(a) + A(a) \stackrel{\mathcal{M}_1}{=} (B + A)(a)$

(ii) $(\varrho + \sigma)(a, b) \overset{\mathscr{R}_1}{=} \varrho(a, b) + \sigma(a, b) = \sigma(a, b) + \varrho(a, b) \overset{\mathscr{R}_1}{=} (\sigma + \varrho)(a, b)$

(iii) $z \cdot (\varrho + \sigma)(a, b) \overset{\mathscr{R}_2}{=} z \cdot ((\varrho + \sigma)(a, b)) \overset{\mathscr{R}_1}{=} z \cdot (\varrho(a, b) + \sigma(a, b)) = z \cdot \varrho(a, b)$
$+ z \cdot \sigma(a, b) \overset{\mathscr{R}_2}{=} (z \cdot \varrho)(a, b) + (z \cdot \sigma)(a, b) \overset{\mathscr{R}_1}{=} ((z \cdot \varrho) + (z \cdot \sigma))(a, b).$

(iv) $\varrho \circ (z \cdot \sigma)(a, b) \overset{\mathscr{R}_5}{=} \sum_{e \in D} \varrho(a, e) \cdot (z \cdot \sigma(e, b)) = z \cdot \sum_{e \in D} \varrho(a, e) \cdot \sigma(e, b)$
$\overset{\mathscr{R}_5}{=} z \cdot (\varrho \circ \sigma)(a, b).$

(v) $(\varrho + \sigma)[a](b) \overset{\mathscr{R}_3}{=} (\varrho + \sigma)(a, b) \overset{\mathscr{R}_1}{=} \varrho(a, b) + \sigma(a, b) \overset{\mathscr{R}_3}{=} \varrho[a](b) + \sigma[a](d)$
$\overset{\mathscr{M}_1}{=} (\varrho[a] + \sigma[a])(b).$

(vi) $\varrho[A + B](a) \overset{\mathscr{R}_4}{=} \sum_{e \in D} (A + B)(e) \cdot \varrho(e, a) \overset{\mathscr{M}_1}{=} \sum_{e \in D} (A(e) + B(e)) \cdot \varrho(e, a)$
$= \left(\sum_{e \in D} A(e) \cdot \varrho(e, a)\right) + \left(\sum_{e \in D} B(e) \cdot \varrho(e, a)\right) \overset{\mathscr{R}_4}{=} \varrho[A](a) + \varrho[B](a)$

(vii) $(\varrho \circ \sigma)[a](b) \overset{\mathscr{R}_3}{=} (\varrho \circ \sigma)(a, b) \overset{\mathscr{R}_5}{=} \sum_{e \in D} \varrho(a, e) \cdot \sigma(e, b)$
$\overset{\mathscr{R}_3}{=} \sum_{e \in D} (\varrho[a](e)) \cdot \sigma(e, b) \overset{\mathscr{R}_4}{=} \sigma[\varrho[a]](b)$

(viii) $(\varrho \circ (\sigma + \tau))(a, b) \overset{\mathscr{R}_5}{=} \sum_{e \in D} \varrho(a, e) \cdot (\sigma + \tau)(e, b)$
$\overset{\mathscr{R}_1}{=} \sum_{e \in D} \varrho(a, e) \cdot (\sigma(e, b) + \tau(e, b)) = \sum_{e \in D} \varrho(a, e) \cdot \sigma(e, b)$
$+ \sum_{e \in D} \varrho(a, e) \cdot \tau(e, b) \overset{\mathscr{R}_5}{=} ((\varrho \circ \sigma) + (\varrho \circ \tau))(a, b).$ □

For the calculus of invariants we shall use, analogously to Chap. 6, a representation of relation nets as matrices and the description of transition firings as vector additions. That is why we now consider vectors consisting of multisets and multirelations, respectively, and show how to calculate with them.

(c) Definition. Let S and D be two sets. Let $X, Y: S \to \mathscr{M}(D)$ be vectors consisting of multisets, and let $\Phi, \Psi: S \to \mathscr{R}(S)$ be multirelation vectors. Let $d \in D$ and let $z \in \mathbb{Z}$. As usual, we define addition and product with integers:

$(\mathscr{V}_1)$ $X + Y\colon S \to \mathscr{M}(D)$, $s \mapsto X(s) + Y(s)$

$(\mathscr{V}_2)$ $z \cdot X\colon S \to \mathscr{M}(D)$, $s \mapsto z \cdot (X(s))$

$(\mathscr{V}_3)$ $\Phi + \Psi\colon S \to \mathscr{R}(D)$, $s \mapsto \Phi(s) + \Psi(s)$

$(\mathscr{V}_4)$ $z \cdot \Phi\colon S \to \mathscr{R}(D)$, $s \mapsto z \cdot (\Phi(s))$

For multirelation vectors, the application to elements of D is defined componentwise:

$(\mathscr{V}_5)$ $\Phi\langle d\rangle\colon S \to \mathscr{M}(D)$, $s \mapsto \Phi(s)[d].$

Finally, we define two operations for multirelation vectors which yield multisets and multirelations, respectively:

the vector application ($\mathscr{V}_6$) $\quad \Phi[X] = \sum_{s \in S} \Phi(s)[X(s)] \quad \in \mathscr{M}(D)$

and the vector product ($\mathscr{V}_7$) $\quad \Phi * \Psi = \sum_{s \in S} \Phi(s) \circ \Psi(s) \quad \in \mathscr{R}(D)$.

The nullary relation vector $\underset{\sim}{O}$ is defined as $\underset{\sim}{O}: S \to \mathscr{R}(D)$

$$s \mapsto O.$$

(d) Lemma. *Let S and D be sets, let $\Phi, \Psi, \Omega: S \to \mathscr{R}(D)$ be vectors and let $z \in \mathbb{Z}$.*

(i) $\Phi * (\Psi + \Omega) = (\Phi * \Psi) + (\Phi * \Omega)$

(ii) $\Phi * (z \cdot \Psi) \quad = z \cdot (\Phi * \Psi)$

(iii) $\Phi[\Psi\langle d\rangle] \quad = (\Psi * \Phi)[d]$.

Proof. (i) $\Phi * (\Psi + \Omega) \overset{\mathscr{V}_7}{=} \sum_{s \in S} \Phi(s) \circ ((\Psi + \Omega)(s)) \overset{\mathscr{V}_3}{=} \sum_{s \in S} \Phi(s) \circ (\Psi(s) + \Omega(s))$

$$\overset{9.4\,(b)\,vii}{=} \sum_{s \in S} (\Phi(s) \circ \Psi(s)) + (\Phi(s) \circ \Omega(s)) \overset{9.4\,(b)\,ii}{=} \sum_{s \in S} \Phi(s) \circ \Psi(s)$$

$$+ \sum_{s \in S} \Phi(s) \circ \Omega(s) \overset{\mathscr{V}_7}{=} (\Phi * \Psi) + (\Phi * \Omega).$$

(ii) $\Phi * (z \cdot \Psi) \overset{\mathscr{V}_7}{=} \sum_{s \in S} \Phi(s) \circ ((z \cdot \Psi)(s)) \overset{\mathscr{V}_4}{=} \sum_{s \in S} \Phi(s) \circ (z \cdot (\Psi(s)))$

$$\overset{9.4\,(b)\,iv}{=} \sum_{s \in S} z \cdot (\Phi(s) \circ \Psi(s)) \overset{9.4\,(b)\,iii}{=} z \cdot \left(\sum_{s \in S} \Phi(s) \circ \Psi(s)\right) \overset{\mathscr{V}_7}{=} z \cdot (\Phi * \Psi).$$

(iii) $\Phi[\Psi\langle d\rangle] \overset{\mathscr{V}_6}{=} \sum_{s \in S} \Phi(s)[\Psi\langle d\rangle(s)] \overset{\mathscr{V}_5}{=} \sum_{s \in S} \Phi(s)[\Psi(s)[d]]$

$$\overset{9.4\,(b)\,vii}{=} \sum_{s \in S} (\Psi(s) \circ \Phi(s))[d] \overset{9.4\,(b)\,v}{=} \left(\sum_{s \in S} \Psi(s) \circ \Phi(s)\right)[d] \overset{\mathscr{V}_7}{=} (\Psi * \Phi)[d].$$ □

9.5 A Matrix Representation for Relation Nets

(a) Definition. Let N be a relation net.

(i) For transitions $t \in T_N$, let the vector $\underline{t}: S_N \to \mathscr{R}(D_N)$ be defined as

$$\underline{t}(s) = \begin{cases} -\overline{(s,t)} & \text{iff } s \in {}^{\bullet}t \setminus t^{\bullet}, \\ \overline{(t,s)} & \text{iff } s \in t^{\bullet} \setminus {}^{\bullet}t, \\ \overline{(t,s)} - \overline{(s,t)} & \text{iff } s \in {}^{\bullet}t \cap t^{\bullet}, \\ O, & \text{otherwise.} \end{cases}$$

(ii) Let the matrix $\underline{N}: S_N \times T_N \to \mathscr{R}(D_N)$ be defined as $\underline{N}(s,t) = \underline{t}(s)$.

(iii) For $\Psi: S_N \to \mathscr{R}(D_N)$, let $\underline{N}' * \Psi: T_N \to \mathscr{R}(D_N)$

$$t \mapsto \underline{t} * \Psi.$$

(b) Theorem. *Let N be a relation net, let $M, M' \in [M_N\rangle$, let $t \in T_N$ and $d \in D_N$. If t is M-enabled with d then $M[t\rangle_d M'$ iff $M + \underline{t}\langle d\rangle = M'$.*

Proof. Let $s \in {}^{\bullet}t \setminus t^{\bullet}$. Then $M'(s) \overset{9.2\,(e)}{=} M(s) - \overline{(s,t)}[d] \overset{9.5\,(a)}{=} M(s) + \underline{t}(s)[d]$ $\overset{\mathscr{V}_5}{=} M(s) + \underline{t}\langle d\rangle(s)$.

	t	u	i_1	i_2	i_3	M_N
d	$-\underline{id}$	$\underline{id}$	$\underline{id}$	$l+r$		$\{p_1,\dots,p_n\}$
g	$-(l+r)$	$l+r$		$-\underline{id}$	$\underline{id}$	$\{g_1,\dots,g_n\}$
e	$\underline{id}$	$-\underline{id}$	$\underline{id}$		$l+r$	$\underline{U}$

Fig. 106. Matrix, invariants and the initial marking of the system shown in Fig. 102

For $s \in t^{\bullet} \setminus {}^{\bullet}t$, $s \in t^{\bullet} \cap {}^{\bullet}t$ and $s \notin t^{\bullet} \cup {}^{\bullet}t$, it can be shown analogously that $M'(s) = M(s) + \underline{t}\langle d\rangle(s)$. The result follows. □

9.6 *S*-Invariants for Relation Nets

(a) Definition. Let N be a relation net. A place vector $i: S_N \to \mathscr{R}(D_N)$ is called an *S-invariant of N* iff $N' * i = \underset{\sim}{O}$.

(b) Corollary. *Let i_1 and i_2 be two S-invariants of a relation net N and let $z \in \mathbb{Z}$. Then $i_1 + i_2$ and $z \cdot i_1$ are also S-invariants of N.*

Proof. Let $t \in T_N$.

(i) $\underline{t} * (i_1 + i_2) \overset{9.4\,(d)\,i}{=} \underline{t} * i_1 + \underline{t} * i_2 \overset{\text{by hypothesis}}{=} O + O = O$.

(ii) $\underline{t} * (z \cdot i_1) \overset{9.4\,(d)\,ii}{=} z \cdot (\underline{t} * i_1) \overset{\text{by hypothesis}}{=} z \cdot O = O$. □

(c) Theorem. *Let N be a relation net. Then, for each S-invariant i of N and each reachable marking $M \in [M_N\rangle$, $i[M] = i[M_N]$.*

Proof. Let $M, M' \in [M_N\rangle$, let $d \in D_N$ and let $t \in T_N$ such that $M[t\rangle_d M'$.

$$i[M'] \overset{9.5\,(b)}{=} i[M + \underline{t}\langle d\rangle] \overset{\mathscr{V}_6}{=} \sum_{s \in S_N} i(s)\,[(M + \underline{t}\langle d\rangle)(s)]$$

$$\overset{\mathscr{V}_1}{=} \sum_{s \in S_N} i(s)\,[M(s) + \underline{t}\langle d\rangle(s)] \overset{9.4\,(b)\,vi}{=} \sum_{s \in S_N} (i(s)\,[M(s)] + i(s)\,[\underline{t}\langle d\rangle(s)])$$

$$\overset{9.4\,(b)\,i}{=} \Big(\sum_{s \in S_N} i(s)\,[M(s)]\Big) + \Big(\sum_{s \in S_N} i(s)\,[\underline{t}\langle d\rangle(s)]\Big) \overset{\mathscr{V}_6}{=} i[M] + i[\underline{t}\langle d\rangle]$$

$$\overset{9.4\,(d)\,iii}{=} i[M] + (\underline{t} * i)\,[d] = i[M] + O[d] = i[M] + \underline{0} = i[M].$$ □

9.7 An Example for Applying *S*-Invariants: The Verification of Facts

Often it is possible to use invariants for proving system properties which are formulated as facts. We shall show this for the facts specified in Fig. 100 in the system of the dining philosophers. Figure 102 shows this system for an arbitrary number, n, of philosophers, represented as a relation net, N. The capacity K_N is given as $K_N(s) = \underline{1}$ for all $s \in S_N$. The matrix and some invariants of the system are shown in Fig. 106. In the following, we use the notation and abbreviations of 9.2 (b).

Proposition. *The T-elements $t_1, \ldots, t_6$ of the system shown in Fig. 102 are facts.*

Proof. The proof is based on seven propositions.

Proposition 1. *Let $M: S_N \to \mathcal{M}(D_N)$ and let $a \in D_N$ such that t_1 is M-enabled with a. Then $M(d) + M(e) \geq 2a$.*

Proof. Using Definition 9.2 (e), we have $M(d) \geq \underline{\mathrm{id}}\,[a]$ and $M(e) \geq \underline{\mathrm{id}}\,[a]$, and hence $M(d) + M(e) \geq 2a$.

Proposition 2. $\forall M \in [M_N\rangle : M(d) + M(e) = P$.

Proof. $M(d) + M(e) = \underline{\mathrm{id}}\,[M(d)] + \underline{\mathrm{id}}\,[M(e)] = i_1[M] = i_1[M_N] = \underline{\mathrm{id}}\,[M_N(d)] + \underline{\mathrm{id}}\,[M_N(e)] = M_N(d) + M_N(e) = P + \underline{0} = P$.

To show that t_1 is a fact, notice that according to Proposition 1, if t_1 is M-enabled with a, then $M(d) + M(e) \geq 2a$. But, using Proposition 2, we have for all $M \in [M_N\rangle : M(d) + M(e) \leq a$. Hence $M \notin [M_N\rangle$ and t_1 is a fact.

Proposition 3. *Let $M: S_N \to \mathcal{M}(D_N)$ and let $a \in D_N$ such that t_2 is M-enabled with a. Then $M(d)(a) + M(e)(a) \leq 0$.*

Proof. Using Definition 9.2 (e), we have $M(d) \leq \underline{1} - a$ and $M(e) \leq \underline{1} - a$, hence $M(d) + M(e) \leq \underline{2} - 2a$, and $M(d)(a) + M(e)(a) \leq (\underline{2} - 2a)(a) = \underline{2}(a) - 2a(a) = 2 - 2 = 0$.

To show that t_2 is a fact, notice that according to Proposition 3, if t_2 is M-enabled with a, then $M(d)(a) + M(e)(a) \leq 0$. Since $^{\bullet}s = \emptyset$, $M(s) \leq M_N(s) = P$. Since $s \in {}^{\bullet}t$ and $\overline{(s, t)} = \underline{\mathrm{id}}$, $a \in P$. But using Proposition 2, for all $M' \in M_N$: $M'(d)(a) + M'(e)(a) = 1$. Hence $M \notin [M_N\rangle$ and t_2 is a fact.

Proposition 4. *Let M be a marking of N and let $a \in D_N$ such that t_3 is M-enabled with a. Then $(l + r)[M(d)] - M(g) \neq G$.*

Proof. Using Definition 9.2 (e), $M(d) \leq \underline{1} - a$ and $M(g) \geq l[a]$, hence $(l + r)[M(d)] \leq (l + r)[\underline{1} - a]$ and $-M(g) \leq -l[a]$. This yields $(l + r)[M(d)] - M(g) \leq (l + r)[\underline{1} - a] - l[a] = (l + r)[\underline{1}] - (l + r)[a] - l[a] = [l + r][\underline{1}] - 2l[a] - r[a] = l[\underline{1}] + r[\underline{1}] - 2l[a] - r[a] = G + G - 2l[a] - r[a] = 2G - 2l[a] - r[a] \neq G$.

Proposition 5. $\forall M \in [M_N\rangle : (l + r)[M(d)] - M(g) = G$.

Proof. $(l+r)[M(d)] - M(g) = (l+r)[M(d)] - \underline{id}[M(g)] = i_2[M] = i_2[M_N] = (l+r)[P] - G = 2G - G = G.$

To show that t_3 is a fact, notice that according to Proposition 4, if t_3 is M-enabled with a, then $(l+r)[M(d)] - M(g) \neq G$. But using Proposition 5, for all $M' \in [M_N\rangle$: $(l+r)[M'(d)] - M'(g) = G$. Hence $M \notin [M_N\rangle$ and t_3 is a fact.

For t_4 the proof is analogous to that for t_3.

Proposition 6. *Let t_5 be M-enabled with a. Then $(l+r)[M(e)] + M(g) \neq G$.*

Proof. Using Definition 9.2, $M(e) \geq a$ and $M(g) \geq l(a)$, hence $(l+r)[M(e)] \geq (l+r)[a]$ and $M(g) \geq l(a)$. This yields $(l+r)[M(e)] + M(g) \geq (l+r)[a] + l[a] = l[a] + r[a] + l[a] = 2l[a] + r[a] \neq G$.

Proposition 7. $\forall M \in [M_N\rangle : (l+r)[M(e)] + M(g) = G.$

Proof. $(l+r)[M(e)] + M(g) = i_3[M] = i_3[M_N] = (l+r)[\underline{0}] + G = G.$

To show that t_5 is a fact, notice that according to Proposition 6, if t_5 is M-enabled with a, then $(l+r)[M(d)] - M(g) \neq G$. Using Proposition 7, $M \notin [M_N\rangle$. Hence t_5 is a fact.

For t_6 the proof is analogous to that for t_5. □

9.8 Relation Net Schemes

In many cases it is possible to derive properties of a relation net without specifying the underlying algebra. These properties then hold for all algebras with

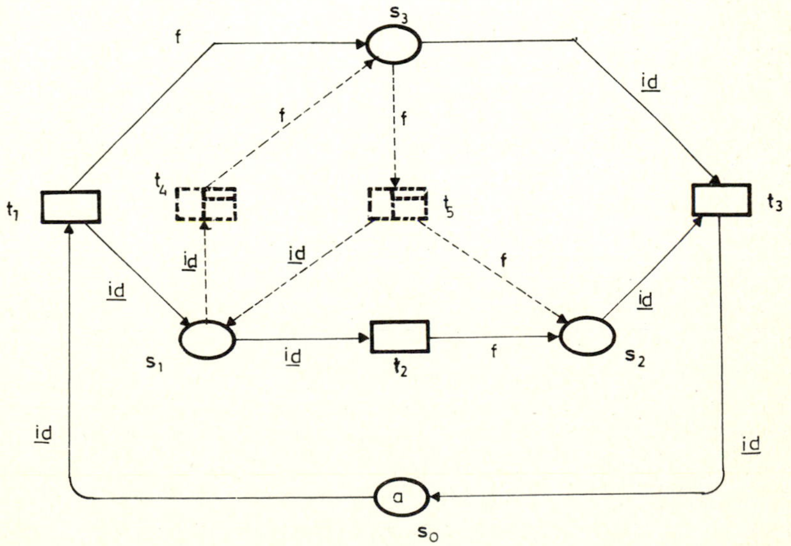

Fig. 107. A net scheme with two facts, t_4 and t_5

	t_1	t_2	t_3	i	M_N
s_0	$-id$		id		a
s_1	id	$-id$		f	
s_2		f	$-id$	id	
s_3	f		$-id$	$-id$	

Fig. 108. Matrix, an invariant and the initial case of the net shown in Fig. 107

corresponding operations, or if additional assumptions are made, for special classes of such algebras. So we now consider relation net schemes which are labelled by element and function symbols instead of concrete elements and functions, respectively.

(a) Figure 107 shows such a relation net scheme with two facts. Indeed, t_4 and t_5 are facts for each concrete interpretation of f and a. We prove this using the invariant which is given in Fig. 108.

Proposition. *The T-elements t_4 and t_5 of the net N shown in Fig. 107 are facts for any algebra $\underline{D}_N = (D; \{f\})$, assuming the capacity $K_N = \underline{1}$.*

Proof. Let $\underline{D}_N (D; \{f\})$ be an arbitrary algebra for N. The proof is based on three propositions.

Proposition 1. *Let $M: S_N \to D_N$ and let $d \in D_N$ such that t_4 is M-enabled with d. Then $f[M(s_1)] + M(s_2) - M(s_3) \neq \underline{0}$.*

Proof. Using Definition 9.2 (e), $M(s_1) \geq d$ and $M(s_3) \leq \underline{1} - f[d]$, hence $f[M(s_1)] + M(s_2) - M(s_3) \geq f[d] + M(s_2) - \underline{1} + f[d] = 2f[d] - \underline{1} + M(s_2) \geq 2f[d] - \underline{1} \neq \underline{0}$.

Proposition 2. $\forall M \in [M_N\rangle$: $f[M(s_1)] + M(s_2) - M(s_3) = \underline{0}$.

Proof. $f[M(s_1)] + M(s_2) - M(s_3) = i[M] = i[M_N] = \underline{0}$.

To show that t_4 is a fact, using Proposition 1 and Proposition 2, we find that t_4 is not enabled for any marking $M \in [M_N\rangle$ and any $d \in D_N$.

Proposition 3. *Let $M: S_N \to D_N$ and let $d \in D_N$ such that t_5 is M-enabled with d. Then $f[M(s_1)] + M(s_2) - M(s_3) \neq \underline{0}$.*

Proof. Using Definition 9.2 (e), $M(s_3) \geq f[a]$, $M(s_1) \leq \underline{1} - a$, $M(s_2) \leq \underline{1} - f[a]$, hence $f[M(s_1)] + M(s_2) - M(s_3) \leq f[\underline{1} - a] + (\underline{1} - f[a]) - f[a] = f[\underline{1}] - f[a] + \underline{1} - f[a] - f[a] = \underline{1} + f[\underline{1}] - 3f[a] \neq \underline{0}$.

To show that t_5 is a fact, using Proposition 2 and Proposition 3, we find that t_5 is not enabled for any marking $M \in [M_N\rangle$ and any $d \in D_N$. □

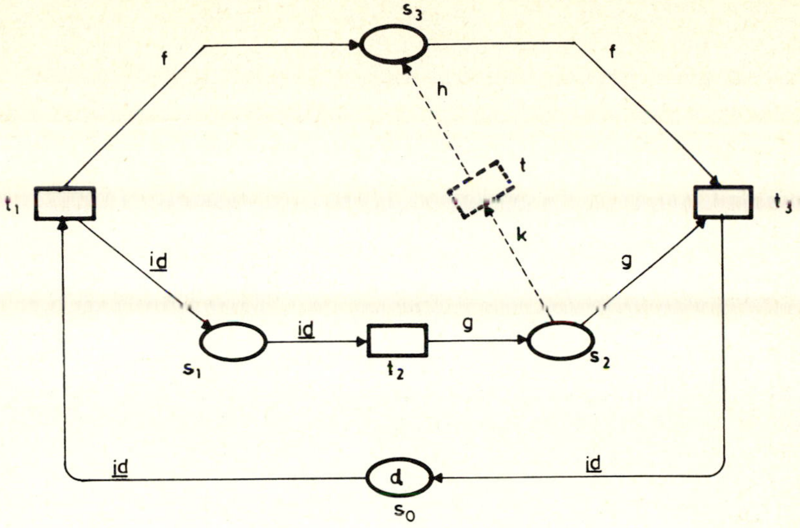

Fig. 109. A net scheme

(b) In the relation net scheme shown in Fig. 109 we assume, for each place $s \in S_N$, the capacity $K_N(s) = \underline{1}$. N contains a T-element t, drawn with broken lines, which is a fact for some but not for all interpretations. For the free algebra, generated by the (unary) operations f, g, h and k, t is certainly a fact: If t could be M-enabled with some a then we would have $k(a) \in M(s_2)$. Since no arc ending at s_2 is labelled with k, this is impossible. t is also a fact if $f = g = h = k$ or, as we shall see later, $k = g \circ g^{-1} = \text{id}$ or $h = g^{-1} \circ f$. t is not a fact if $k = g^{-1}$ and $h \neq f$, because t would then be enabled with d for the marking $(\underline{0}, \underline{0}, g[d], f[d])$ (a marking M of N is here represented as $(M(s_0), \ldots, M(s_3))$.

	t_1	t_2	t_3	i_1	i_2	i_3	i_4	i_5	M_N
s_0	$-\underline{id}$		$\underline{id}$	g	f				d
s_1	$\underline{id}$	$-\underline{id}$		g		$\underline{id}$	$f \circ g$	f	
s_2		g	$-g$	$\underline{id}$		g^{-1}	f	$g^{-1} \circ f$	
s_3	f		$-f$		$\underline{id}$	$-f^{-1}$	$-g$	$-\underline{id}$	
						iff	iff	iff	
						$f \circ f^{-1} = \underline{id} = g \circ g^{-1}$	$f \circ g = g \circ f$	$g \circ g^{-1} = \underline{id}$	

Fig. 110. Matrix, invariants and the initial case of the net shown in Fig. 109

Accordingly, we find for the invariants of N (see Fig. 110) that only i_1 and i_2 are invariants under all interpretations. The vectors i_3 to i_5 are only invariants if the interpretation fulfils the respective conditions.

Again, we can use the knowledge about invariants to verify facts. We show that t is a fact for all interpretations with $k = g \circ g^{-1} = \underline{\mathrm{id}}$ and $h = g^{-1} \circ f$.

Proposition 1. *Let $M: S_N \to D_N$ and let $d \in D_N$ such that t is M-enabled with d. Then $f[M(s_1)] + h[M(s_2)] - M(s_3) \neq \underline{0}$.*

Proof. Using Definition 9.2 (*a*), $M(s_2) \geq \underline{\mathrm{id}}\,[d] \wedge M(s_3) \leq \underline{1} - h[d]$, i.e. $M(s_2) \geq d \wedge -M(s_3) \geq h[d] - \underline{1}$. This yields $h[M(s_2)] \geq h[d] \wedge -M(s_3) \geq h[d] - \underline{1}$. Hence $h[M(s_2)] - M(s_3) \geq 2h[d] - \underline{1}$ and, since $M(s_1) \geq \underline{0}$, $f[M(s_1)] + h[M(s_2)] - M(s_3) \geq 2h[d] - \underline{1}$. In particular we have $(2h[d] - \underline{1})(h[d]) = 2h[d] - h[d] = h[d]$ and the result follows. □

Proposition 2. $\forall M \in [M_N\rangle : f[M(s_1)] + h[M(s_2)] - M(s_3) = \underline{0}$.

Proof. $f[M(s_1)] + h[M(s_2)] - M(s_3) = f[M(s_1)] + g^{-1} \circ f[M(s_2)] - \underline{\mathrm{id}}\,[M(s_3)] = i[M] = i[M_N] = \underline{0}$. □

Using Proposition 1 and 2, it follows immediately that t is not enabled for any marking $M \in [M_N\rangle$ and any $d \in D_N$.

Appendix

Mathematical Notions and Notation

I. Sets

A1. As usual we use, for sets M, the notation $x \in M$ and $A \subseteq M$ to denote that x is an element of M and A is a subset of M. $\mathscr{P}(M)$ denotes the powerset of M.

A2. Let A, B, C be sets. As usual, $A \cup B$, $A \cap B$ and $A \backslash B = \{a \in A \mid \neg(a \in B)\}$ denote the union of A and B, the intersection of A and B and the complement of B in A.

A3. From set theory, we use the distributive laws $A \cap (B \cup C) = (A \cap B) \cup (A \cap C)$, $A \cup (B \cap C) = (A \cup B) \cap (A \cup C)$ and

(i) $A\backslash(B \cup C) = (A\backslash B)\backslash C$,
(ii) $A\backslash(A\backslash B) = A \cap B$,
(iii) $(A \cup B)\backslash C = (A\backslash C) \cup (B\backslash C)$,
(iv) $(A\backslash B)\backslash A = \emptyset$,
(v) $A\backslash(B \cup C) = (A\backslash B) \cap (A\backslash C)$,
(vi) $(A\backslash B) \cap C = (A \cap C)\backslash(B \cap C)$,
(vii) $A \subseteq B \Rightarrow A\backslash B = \emptyset$.

A4. $\mathbb{N}$ denotes the set of natural numbers $\{0, 1, 2, \ldots\}$ and $\mathbb{Z}$ denotes the set of integers $\{\ldots -2, -1, 0, 1, 2, \ldots\}$.

II. Relations

A5. Definition. Let M be a set. For $x, y \in M$, (x, y) is called a *pair* over M. For $A, B \subseteq M$, let $A \times B = \{(x, y) \mid x \in A \wedge y \in B\}$. $\varrho \subseteq M \times M$ is called a *relation* and we write $x \varrho y$ for $(x, y) \in \varrho$.

A6. Definition. Let M be a set and let $\varrho, \sigma \subseteq M \times M$ be two relations over M. We define:

(i) $\varrho^{-1} = \{(y, x) \mid (x, y) \in \varrho\}$.
(ii) $\varrho \circ \sigma = \{(x, z) \mid \exists y \in M \; x \varrho y \wedge y \sigma z\}$.
(iii) With $\varrho^0 = \{(x, x) \mid x \in M\}$ and $\varrho^{i+1} = \varrho^i \circ \varrho$ $(i = 0, 1, \ldots)$, let $\varrho^+ = \bigcup_{i=1}^{\infty} \varrho^i$ and $\varrho^* = \varrho^+ \cup \varrho^0$.
(iv) For $a \in M$, let $\varrho[a] = \{b \in M \mid a \varrho b\}$.

A7. Corollary. *If $\varrho, \sigma \subseteq M \times M$ are relations then*
(i) $\varrho = \varrho^1$,
(ii) $\varrho \subseteq \sigma \Rightarrow \varrho^* \subseteq \sigma^*$,
(iii) $\varrho^* \cup \sigma^* \subseteq (\varrho \cup \sigma)^*$,
(iv) $(\varrho^*)^* = \varrho^*$.

A8. Lemma. *Let $\varrho, \sigma, \tau, \psi \subseteq M \times M$ be relations. Then*
(i) $\varrho \subseteq \sigma^* \wedge \tau \subseteq \psi^* \Rightarrow (\varrho \cup \tau)^* \subseteq (\sigma \cup \psi)^*$
(ii) $\varrho \subseteq \sigma^* \Rightarrow (\varrho \cup \sigma)^* \subseteq \sigma^*$.

Proof. (i) $\varrho \cup \tau \subseteq \sigma^* \cup \psi^* \subseteq (\sigma \cup \psi)^* \Rightarrow (\varrho \cup \tau)^* \subseteq ((\sigma \cup \psi)^*)^* = (\sigma \cup \psi)^*$.
(ii) $\varrho \subseteq \sigma^* \Rightarrow \varrho \cup \sigma^* \subseteq \sigma^* \Rightarrow (\varrho \cup \sigma^*)^* \subseteq (\sigma^*)^* = \sigma^* \Rightarrow (\varrho \cup \sigma)^* \subseteq \sigma^*$. □

III. Mappings, Functions

A9. Definition. Let A, B be sets and let $M \subseteq A$.
(i) $f: A \to B$ denotes a (total) function (or mapping) from A to B.
(ii) For $f: A \to B$, let $f(M) = \{f(a) \mid a \in M\}$.
(iii) The mapping $f \mid M: M \to B$ is defined as $f \mid M(a) = f(a)$ for all a $\in M$.
(iv) The relation $\{(a, f(a)) \mid a \in A\}$ is called the *graph* of the function $f: A \to B$.

A10. Definition. Let A be a set.
(i) $\underline{\text{id}}: A \to A$ with $\underline{\text{id}}(a) = a$ is called the *identity function* or *identity*.
(ii) For $n, i \in \mathbb{N}$ let $\underline{\text{pr}}_i: A^n \to A$ be defined by $\underline{\text{pr}}_i(a_1, \ldots, a_n) = a_i$.

IV. Partial Orders

A11. Definition. Let M be a set. A relation $\varrho \subseteq M \times M$ is called a *partial order* iff $\forall a, b \in M$:
(i) $\neg(a \,\varrho\, a)$ (ϱ is irreflexive),
(ii) $a \,\varrho\, b \wedge b \,\varrho\, c \Rightarrow a \,\varrho\, c$ (ϱ is transitive).

Note that (i) and (ii) imply the asymmetry of ϱ: $a \,\varrho\, b \Rightarrow \neg(b \,\varrho\, a)$.

Without regard to the carrier, we write partial orders $\varrho \subseteq M \times M$ as "$<$". Let $a \leq b \Leftrightarrow a < b \vee a = b$.

Graphically, we present finite partial orders as graphs such that an arc $a \to b$ is drawn iff $a < b \wedge \nexists c: a < c < b$.

V. Graphs

A12. Definition. A tuple $G = (H, P)$ is called an *(arc labelled, oriented) graph over* L iff H and L are sets such that $P \subseteq H \times L \times H$. The elements of H, L and P are called *nodes, arc labels* and *arcs,* respectively.

The graphical representation of graphs is obvious.

A13. Definition. Let $G = (H, P)$ be a graph over L. For $i = 1, 2, \ldots$ Let $p_i = (h_i, l_i, h_i') \in P$. $w = p_1 p_2 \ldots$ is called a *path* in G iff, for $i = 1, 2, \ldots, h_i' = h_{i+1}$. Then we also write $w = h_1\, l_1\, h_2\, l_2 \ldots$. w is *finite* iff for some $n \in \mathbb{N}$, p_{n+1} is not constructed. In this case, n is the *length* of w. The *empty path* ε is of length 0. w is a *circle* iff, for some $n \in \mathbb{N}$, w is of length n and $h_n = h_1$.

A14. Definition. Let $G_i = (H_i, P_i)$ be graphs over L_i $(i = 1, 2)$. G_1 is called *α–β-isomorphic* (*isomorphic*, for short) to G_2 iff $\alpha \colon H_1 \to H_2$ and $\beta \colon L_1 \to L_2$ are bijective mappings such that $(h, l, h') \in P_1 \Leftrightarrow (\alpha(h), \beta(l), \alpha(h')) \in P_2$.

A15. Definition. Let $G = (H, P)$ be a graph.

(i) G is *acyclic* iff G contains no circles.
(ii) $h \in H$ is an *initial node* iff $\{(h_1, l, h_2) \in P \mid h_2 = h\} = \emptyset$.
(iii) G is *finitely based* iff G has only finitely many initial nodes.
(iv) G is *finitely branched* iff for each node $h \in H$, $\{(h_1, l, h_2) \in P \mid h_1 = h\}$ is finite.

We state now the well known Lemma of König in a form which is appropriate for our purposes:

A16. Theorem. *Let $G = (H, P)$ be an acyclic, finitely based and finitely branched graph. If every path of G is finite then G itself is finite.*

Proof. For $h \in H$, let suc (h) be the set of nodes $h' \in H$ such that there exists a path from h to h'.

Assume G is infinite. We construct an infinite path $h_1\, l_1\, h_2\, l_2 \ldots$ as follows: As G is finitely based there exists at least one initial node h such that suc (h) is infinite. Let $h_1 = h$. By induction assume h_i being given, and let suc (h_i) be infinite. As h_i is finitely branched, there exists at least one arc (h_i, l, h') such that suc (h') is infinite. Then let $l_i = l$ and $h_{i+1} = h'$. □

VI. Suprema of Sets of Natural Numbers and Calculation with ω

A17. Definition. (i) We expand the canonical ordering $<$ and the operations $+$ and $-$ on $\mathbb{N}$ to $\mathbb{N} \cup \{\omega\}$ such that $\forall n \in \mathbb{N} \colon n < \omega$ and $\forall m \in \mathbb{N} \cup \{\omega\}$: $m + \omega = \omega + m = \omega$; $\omega - m = \omega$.

(ii) For $A \subseteq \mathbb{N} \cup \{\omega\}$, let

$$\underline{\sup}(A) = \begin{cases} a & \text{iff} \quad a \in A \wedge \forall a' \in A : a' \leq a, \\ \omega & \text{iff} \quad \forall n \in \mathbb{N}\ \exists a \in A : n \leq a. \end{cases}$$

A18. Corollary. *Let $A, B \subseteq \mathbb{N} \cup \{\omega\}$ with $A = \{a_1, a_2, \ldots,\}$ and $B = \{b_1, b_2, \ldots\}$. If $a_1 < b_1 \wedge a_2 < b_2 \wedge \ldots$, then $\underline{\sup}(A) \leq \underline{\sup}(B)$.*

VII. Vectors and Matrices

We shall use arbitrary finite sets to index vectors and matrices (instead of the more usual sequences of natural numbers). The components will be integers.

A19. Definition. Let A be a non-empty, finite set. A mapping $v: A \to \mathbb{Z}$ is called a *vector* or an *A-vector*. For two vectors $v_1 : A \to \mathbb{Z}$ and $v_2 : A \to \mathbb{Z}$, we define

(i) their *sum* $v_1 + v_2$ as the vector $v: A \to \mathbb{Z}$ with $v(a) = v_1(a) + v_2(a)$,

(ii) their *product* $v_1 \cdot v_2$ as the integer $\sum_{a \in A} v_1(a) \cdot v_2(a)$,

(iii) for $z \in \mathbb{Z}$, the *scalar product* $z \cdot v_1$ as the vector $v: A \to \mathbb{Z}$ with $v(a) = z \cdot v_1(a)$.

A20. Definition. Let A be a set.

(i) A vector $v: A \to \{0\}$ is called the *null vector* and is denoted by 0 (its domain A is given by the particular context).

(ii) A vector $v: A \to \{0, 1\}$ is called *characteristic*.
For $A' \subseteq A$ let $c_{A'} : A \to \{0, 1\}$,

$$a \mapsto 1 \quad \text{iff} \quad a \in A',$$
$$a \mapsto 0 \quad \text{otherwise}.$$

$c_{A'}$ is called the *characteristic vector of* A'.

(iii) A vector $v: A \to \mathbb{Z}$ is called *positive* iff $\forall a \in A: v(a) \geq 0$.

A21. Definition. Let A and B be non-empty, finite sets which are disjoint.

(i) A mapping $C: A \times B \to \mathbb{Z}$ is called a *matrix*.

(ii) The *transposed matrix* C' of a matrix $C: A \times B \to \mathbb{Z}$ is the matrix C': $B \times A \to \mathbb{Z}$ with $C'(b, a) = C(a, b)$.

(iii) The *product* of a matrix $C: A \times B \to \mathbb{Z}$ with a vector $v: B \to \mathbb{Z}$ yields the vector $C \cdot v : A \to \mathbb{Z}$ with $C \cdot v(a) = \sum_{b \in B} C(a, b) \cdot v(b)$.

Graphically, vectors and matrices are represented as tables, following the scheme shown in Fig. 111. With $A = \{a_1, \ldots, a_n\}$ and $B = \{b_1, \ldots, b_m\}$, let $v: A \to \mathbb{Z}$ be a vector and $C: A \times B \to \mathbb{Z}$ be a matrix. For $i = 1, \ldots, n$ and $j = 1, \ldots, m$, let $v_i = v(a_i)$ and $C_{ij} = C(a_i, b_j)$.

	v
a_1	v_1
a_2	v_2
⋮	⋮
a_n	v_n

C	b_1	---	b_m
a_1	c_{11}	---	c_{1m}
a_2	c_{21}	---	c_{2m}
⋮	⋮		
a_n	c_{n1}	---	c_{nm}

Fig. 111. Graphical representation of vectors and matrices

Further Reading

We start with a very brief review on the development of Net Theory. Then we survey other text books and mention detailed bibliographies on nets.

Separately for each chapter we will mention a selection of papers which

- are sources of the material presented in this book
- have been the very first ones in the field
- might be considered as typical
- have recently been published and might be a formal basis for further studies.

Finally, we mention modifications and generalizations of place/transition-nets, survey applications and implementations of nets and outline related system models.

1. Some Landmarks in the Development of Net Theory

As already mentioned in the preface, Net Theory started in the early 60ies with the dissertation of C. A. Petri, where the need for a theory of asynchronous machine models is stated:

[1] C. A. Petri: *Kommunikation mit Automaten.* Schriften des Institutes für Instrumentelle Mathematik, Bonn 1962.

There is also an English Translation by Clifford F. Greene, Jr.:

[2] C. A. Petri: *Communication with Automata.* Final report, Volume 1, Supplement 1 RADC TR-65-377-vol-1-suppl 1, Applied Data Research, Princeton, NJ, Contract AF 30 (602)-3324 (January 1966).

Further early publications include:

[3] C. A. Petri: *Fundamentals of a Theory of Asynchronous Information Flow.* Information Processing 1962, Proceedings of the IFIP Congress 62, Munich. North Holland Publishing Company Amsterdam (1962) pp. 386 – 390

and

[4] C. A. Petri: *Grundsätzliches zur Beschreibung diskreter Prozesse.* Drittes Colloquium über Automatentheorie, Birkhäuser Verlag Basel (1967), pp. 121 – 140.

The late sixties saw the Information System Theory Project which dealt with nets of conditions and events:

[5] A. W. Holt, H. Saint, R. Shapiro, S. Warshall: *Final Report of the Information Systems Theory Project.* Technical Report RADC-TR-68-305, Rome Air Development Center, Griffis Air Force Base, New York (Sept. 1968), 352 pages. Distributed by Clearinghouse for Federal Scientific and Technical Information, US Department of Commerce.

In this context, a basic paper is also

[6] S. S. Patil: *Coordination of Asynchronous Events.* PhD Thesis (May 1970) Cambridge, Mass.: MIT Project MAC, Technical Report 72 (June 1970).

which was the beginning of MIT's short involvement in net theory.

As classical examples of papers on place/transition-nets we suggest M. Hack's introduction of free choice nets:

[7] M. H. T. Hack: *Analysis of Production Schemata by Petri Nets.* Technical Report 94, Project MAC (February 1972).

and Commoner's investigations on liveness for arc weighted free choice nets and simple nets:

[8] Frederic G. Commoner: *Deadlocks in Petri Nets.* Applied Data Research Inc., Wakefield, Massachusetts 01880. Report Nr. CA-7206-2311 (1972).

The concepts of processes and of *K*-density have their origin in

[9] C. A. Petri: *Non-sequential Processes.* Internal Report GMD-ISF-77-5 (1977), Gesellschaft für Mathematik und Datenverarbeitung, Bonn.

Nets with individual tokens were introduced by H. Genrich and K. Lautenbach in the following paper:

[10] H. Genrich, K. Lautenbach: *System Modelling with High Level Petri Nets.* Theoretical Computer Science 13 (1981), pp. 109–136.

General Net Theory studies (beside others) concepts to systematically relate the various net models (we did not stress these aspects in this book, but gave a small example in Figs. 9 and 10). C. A. Petri (and others) developed this theory in various papers, e.g.

[11] C. A. Petri: *Concepts of Net Theory.* Mathematical Foundations of Computer Science, Proceedings of Symposium and Summer School, High Tatras, September 3–8, 1973. Math. Inst. of the Slovak Acad. of Science (1973), pp. 137–146

[12] C. A. Petri: *General Net Theory.* Proceedings of the Joint IBM University of Newcastle upon Tyne Seminar on Computing System Design, Sept. 1976, B. Shaw (ed.) (1977).

A recent contribution to the basic ideas of Net Theory is

[13] C. A. Petri: *State-Transition Structures in Physics and in Computation.* International Jorunal of Theoretical Physics, Vol. 21 Nos. 10/11 (1982).

2. Conferences on Petri Nets

The earliest conferences dealing – at least to some extent – with Petri Nets were the MAC Conference on concurrent systems in 1970, the GMD Conference "Ansätze zur Organisation rechnergestützter Informationssysteme" in 1974, the MIT Conference on Petri Nets and Related Methods, 1975 (unpublished) and the Journées d'étude AFCET Réseaux de Petri, 1977.

[14] J. Dennis (Editor): *Record of the Project MAC Conference on Concurrent Systems and Parallel Computation,* New York: AMC (June 1970)

[15] C. A. Petri (Editor): *Ansätze zur Organisationstheorie rechnergestützter Informationssysteme.* R. Oldenbourg Verlag München, Wien. Berichte der Gesellschaft für Mathematik und Datenverarbeitung Nr. 111 (1979)

[16] Institute de Programmation, Universite Paris VI (Editor): *Journées d'étude AFCET Réseaux de Petri,* Paris (1977)

An important event was the Advanced Course on General Net Theory of Processes and Systems in Hamburg, Oct., 8–19, 1979. The course material was published in

[17] W. Brauer (ed.): *Net Theory and Applications.* Springer Lecture Notes in Computer Science, 84 (1980).

This represents the state-of-the-art until about 1979. More recent material is collected in the proceedings of the European Workshops on Application and Theory of Petri Nets:

[18] C. Girault, W. Reisig (eds.): *Application and Theory of Petri Nets.* Informatik Fachbericht 52, Springer Publishing Company (1982)

and

[19] A. Pagnoni, G. Rozenberg (eds.): *Application and Theory of Petri Nets.* Informatik Fachberichte 66, Springer Publishing Company (1983).

3. Text Books

Until recently, there did not exist any text books on Petri Nets. As a substitute, the proceedings [17] have sometime been used as an introductory text. In particular, this volume contains a proposal for a standard terminology which we observed in this book:

[20] H. J. Genrich, E. Stankiewicz-Wiechno: *A Dictionary of Some Basic Notions of Net Theory,* in [17].

In the following we refer to books which are distributed by professional publishers. The many introductory texts in journals or internal reports are not mentioned here.

in English:

[21] J. L. Peterson: *Petri Net Theory and the Modeling of Systems.* Prentice-Hall, Inc., Englewood Cliffs, N.J. 07632 ISMN 0-13-661983-5 (1981)

in French:

[22] G. W. Brams (nom collective): *Réseaux de Petri, Théorie et Pratique.* Masson, Editeur, 120 boulevard Saint-Germain 75280 Paris Cedex 06 ISMN 2-903-60712-5 (1982). Two volumes

in German:

[23] P. H. Starke: *Petri-Netze.* VeB Deutscher Verlag der Wissenschaften, Berlin (DDR) (1981)

[24] U. Winand, B. Rosenstengel: *Petri-Netze. Eine anwendungsorientierte Einführung.* Vieweg-Verlag Braunschweig. ISBN 3-528-03582-X (1981).

The original german version of the present book is published by Springer Verlag. An Italian translation is published by Arnoldo Mondadori Editore, Milano (Italy).

All four books [21] to [24] concentrate on nets consisting of places and transitions.

4. Bibliographies

Many papers are referenced in the various contributions of [17]. A detailed and annotated bibliography, covering papers until 1979, is contained in the book of Peterson [21]. The bibliography

[25] E. Pless, H. Plünnecke: *A Bibliography of Net Theory.* Second Edition ISF-Report 80.05. Gesellschaft für Mathematik und Datenverarbeitung Bonn, Germany (1980)

reports about 500 papers which were published up to 1980. More recent references are continuously published in the newsletter of the GI-Special Interest Group on Petri Nets and Related System Models:

[26] *Newsletter of the Special Interest Group "Petri Nets and Related System Models".* Gesellschaft für Informatik (Computer Science Society in Germany), Bonn, Germany. ISSN 0173-7473.

5. References to Chapter 2

Conditions and events have been fundamental notions of Net Theory from the very beginning. The first extensive studies were published in [5], and can also be found in

[27] A. Holt: *Introduction to Occurrence Systems.* Associative Information Techniques, New York: American Elsevier (1971), pp. 175–203.

For a further early study see also [6].

The notation we use is based on the following two papers:

[28] C. A. Petri: *Interpretations of Net Theory.* Internal Report 75-07, second edition, 20. 12. 1976. Gesellschaft für Mathematik und Datenverarbeitung, Institut für Informationssystemforschung, Bonn (1976)

[29] H. J. Genrich, K. Lautenbach, P. S. Thiagarajan: *Elements of General Net Theory,* in [17].

6. References to Chapter 3

The idea of unfolding a condition/event-system to partially ordered event occurrences was introduced in [5].

The notion of a process, as defined in 3.3 (a), was first discussed by C. A. Petri in [9].

Petri introduces a lot of properties which a "reasonable" notion of process should meet. In [28] a collection of five such properties is chosen to define this notion. The theorems which we prove in Chapt. 3.3 to 3.5 are not given in the literature.

An early paper on K-density is

[30] E. Best: *A Theorem on the characteristics of non-sequential processes.* Fundamenta Informaticae III.1 (1980), pp. 77–94.

More recently, occurrence nets have been studied independently of any correspondence to condition/event systems. As examples see

[31] E. Best, A. Merceron: *Discreteness, K-density and D-continuity of Occurrence Nets.* 6th GI Conference on Theoretical Computer Science. Lecture Notes in Computer Science 145, Springer-Verlag (1983)

[32] C. Fernández, P. S. Thiagarajan: *D-Continuous Causal Nets: A Model of Non-Sequential Processes.* Theoretical Computer Science 28 (1984), pp. 171–196.

In the context of schemes for nonsequential systems, the following papers describe processes with the use of partial orders:

[33] A. Mazurkiewicz: *Concurrent Program Schemes and their Interpretation.* University of Aarhus, DAIMI PB-78 (1978)

[34] J. Winkowski: *Behaviours of Concurrent Systems.* Theoretical Computer Science 12 (1980), pp. 39–60

[35] W. Reisig: *Schemes for Nonsequential Processing Systems.* 9th Symposium on Mathematical Foundations of Computer Science, Lecture Notes in Computer Science 88, Springer-Verlag (1980)

[36] M. Nielsen, G. Plotkin, G. Winskel: *Petri Nets, Event Structures and Domains, Part I.* Theoretical Computer Science 13 (1981), pp. 85–108

[37] J. Winkowski: *An Algebraic Description of System Behaviours.* Theoretical Computer Science 21 (1982), pp. 315–340

[38] G. Winskel: *Events in Computation.* Ph. D. thesis, University of Edinburgh (1980).

7. References to Chapter 4

Synchronic distance was first mentioned in

[39] C. A. Petri: *Concepts of Net Theory.* Mathematical Foundations of Computer Science, 1973. High Tatra; Mathematics Institute of Slovak Academy of Science (1973), pp. 137–146.

There have been some formal definitions, e.g. in [28], but there are some problems in the case of non-cyclic systems. In [29] a definition for synchronic distance is given which is equivalent to ours. More on synchronic distances can be found in

[40] C. André, P. Armand, F. Boeri: *Synchronic Relations and Applications in Parallel Computation.* Digital Processes 5 (1979), pp. 339–351

[41] U. Goltz, W. Reisig, P. S. Thiagarajan: *Two Alternative Definitions of Synchronic Distance,* in [18].

The extension to weighted synchronic distances is discussed in [29] and in

[42] U. Goltz, W. Reisig: *Weighted Synchronic Distances,* in [18].

A typical application of synchronic distances is

[43] A. C. Pagnoni: *A Fair Competition Between Two or More Partners,* in [18].

The idea of facts was first mentioned in [28]. Further investigations on facts can be found in

[44] H. J. Genrich, G. Thieler-Mevissen: *The Calculus of Facts.* Mathematical Foundations of Computer Science 1976, Lecture Notes in Computer Science 45. Springer-Verlag (1976), pp. 588–595

and in

[45] G. Thieler-Mevissen: *The Petri Net Calculus of Predicate Logic.* Internal Report ISF-76-09 (1976), Gesellschaft für Mathematik und Datenverarbeitung, Bonn.

8. References to Chapter 5

To a large extent, papers on Petri Nets deal with place/transition-nets. Indeed, often both notions are synonymously used.

In order to give a representative survey over the area, we subdivide this section into several sub-sections.

(a) Coverability Graphs

The first idea resembling the coverability graphs was the introduction of a "reachability tree" by Karp and Miller:

[46] R. M. Karp, R. E. Miller: *Parallel Program Schemata.* Journal of Computer and System Sciences 3 (1969), pp. 147–195.

A construction method for coverability graphs which differs slightly from ours is presented in

[47] M. Jantzen, R. Valk: *Formal Properties of Place/Transition Nets,* in [17].

(b) Liveness

The notion of liveness has often been considered as a mayor problem for analysis. There exist different reasonable notions of liveness, cf.

[48] K. Lautenbach: *Liveness in Petri Nets.* Internal Report GMD-ISF 72-02.1 (1972).

Papers on liveness include

[49] M. Hack: *The Recursive Equivalence of the Reachability Problem and the Liveness Problem for Petri Nets and Vector Addition Systems.* Proceedings of the 15th Annual Symposium on Switching and Automata Theory, New York IEEE (1974)

[50] K. Gostelow: *Computation Modules and Petri Nets.* Third IEEE-ACM Milwaukee Symposium on Automatic Computation and Control, New York (1975)

[51] H. Schmid, E. Best: *Towards a Constructive Solution of the Liveness Problem in Petri Nets.* Technical Report 4/76, Institut für Informatik, Universität Stuttgart, West Germany (1976)

[52] Y. Lien: *Termination Properties of Generalized Petri Nets.* SIAM Journal of Computing 5, Nr. 2 (1976), pp. 251–265.

Liveness is also discussed in [8] and [5].

(c) Further Properties

Further properties of place/transition nets which have not been discussed in this book include persistence, the existence of homestates and equivalence. Homestates are considered e.g. in the first volume of [22]. Persistency is discussed in the following papers:

[53] L. Landweber, E. Robertson: *Properties of Conflict-Free and Persistent Petri Nets.* Journal of the ACM, Vol. 25, Nr. 3 (1978), pp. 352–364

[54] J. H. Müller: *Decidability of Reachability in Persistent Vector Replacement Systems.* 9th Symposium on Mathematical Foundations of Computer Science, Lecture Notes in Computer Science Vol. 88, Springer-Verlag (1980), pp. 426–438

[55] E. Mayr: *Persistence of Vector Replacement Systems is Decidable.* Acta Informatica 15 (1981), pp. 309–318

[56] H. Yomoasaki: *On Weak Persistency of Petri Nets.* Information Processing Letters 13, 3 (1981), pp. 94–97.

The concept of equivalence is discussed in

[57] J. R. Jump, P. S. Thiagarajan: *On the Equivalence of Asynchronous Control Structures.* 13th Annual Switching and Automata Theory Symposium (Oct. 1972), 212–223. Also: SIAM Journal of Computing, Vol. 2, No. 2 (June 1973), pp. 67–87

[58] C. André: *Use of Behaviour Equivalence in Place/Transition Net Analysis,* in [18].

[59] C. André: *Structural Transformations giving B-equivalent PT-Nets,* in [19]

[60] F. De Cindio, G. De Michelis, L. Pomello, C. Simone: *Equivalence Notions for Concurrent Systems,* in [19]

[61] M. Yoeli, T. Etzion: *Behavioural Equivalence of Concurrent Systems,* in [19].

(d) Analysis Methods

An analysis method for place/transition-nets which we did not mention in this book, is the *reduction* of nets. In the first volume of [22], this method is discussed in detail. It is also presented in

[62] G. Berthelot, G. Roucairol, R. Valk: *Reduction of Nets and Parallel Programs,* in [17].

Similar methods are presented in the following papers:

[63] J. R. Valette: *Analysis of Petri Nets by Stepwise Refinements.* Journal of Computer and System Sciences 18, No. 1 (1979), pp. 35–46

[64] M. Toulotte, J. P. Parsy: *A Method for Decomposing Interpreted Petri Nets and its Utilization.* Digital Processes 5 (1979), pp. 223–234

[65] I. Suzuki, T. Murata: *A Method for Hierarchically Representing Large Scale Petri Nets.* Proceedings of the 1980 International Conference on Circuits and Computer, October 1980

[66] M. Silva: *Simplification des Rèseaux de Petri par elimination des places implicites.* Digital Processes 6 (1980), pp. 245–256.

(e) The Reachability Problem

The reachability problem has been open since the introduction of vector addition systems [46] and was solved recently by Kosaraju:

[67] S. R. Kosaraju: *Decidability of Reachability in Vector Addition Systems.* Proceedings of the Fourteenth Annual ACM Symposium on Theory of Computing, San Francisco, California, May 5–7, 1982, pp. 267–281.

Some corrections of this proof are given in

[68] H. J. Müller: *Filling a Gap in Kosaraju's Proof for the Decidability of the Reachability Problem in VAS.* Newsletter of the Special Interest Group "Petri Nets and Related System Models", No. 12, October 1982 (cf. [26]).

Milestones on the way to this solution are [54] and

[69] J. van Leeuwen: *A Partial Solution to the Reachability Problem for Vector-Addition Systems.* Proc. of the sixth Annual ACM Symposium on Theory of Computing (1974), pp. 303–307.

[70] G. S. Sacerdote, R. L. Tenney: *The Decidability of the Reachability Problem for Vector Addition Systems.* Proc. of the ninth Annual ACM Symposium on Theory of Computing (1977), pp. 61–76.

[71] J. Hopcroft, J. J. Pansiot: *On the Reachability Problem of five dimensional Vector Addition Systems.* Theoretical Computer Science 8 (1979), pp. 135–159.

[72] E. W. Mayr: *An Algorithm for the General Petri Net Reachability Problem.* Proc. of the 13th Annual ACM Symposium on Theory of Computing (1981), pp. 238–246.

(f) Decidability and Complexity

A survey on decidability and complexity problems is contained in

[73] M. Jantzen: *Komplexität von Petrinetz-Algorithmen.* Unpublished course material, University of Hamburg (1984).

The decidability and the complexity of net properties are treated in [21, 47] and in the following papers:

[74] H. Baker: *Rabin's Proof of the Undecidability of the Reachability Set Inclusion Problem for Vector Addition Systems.* Computation Structures Group Memo 79, Project MAC, MIT Cambridge, Massachusetts (July 1973)

[75] M. Hack: *Decidability Question for Petri Nets.* Ph. D. thesis, Department of Electrical Engineering, MIT (December 1974). Also: Technical Report 161, Laboratory for Computer Science, MIT, Cambridge, Massachusetts (June 1976)

[76] M. Hack: *The Equality Problem for Vector Addition Systems is Undecidable.* Theoretical Computer Science 2 (1976), pp. 77–95

[77] T. Araki, T. Kasami: *Some Decision Problems Related to the Reachability Problem for Petri Nets.* Theoretical Computer Science 3 (1977), pp. 85–104

[78] T. Araki, T. Kasami: *Decidable Problems on the Strong Connectivity of Petri Net Reachability Sets.* Theoretical Computer Science 4 (1977), pp. 99–119

[79] N. Jones, L. Landweber, Y. E. Lien: *Complexity of Some Problems in Petri Nets.* Theoretical Computer Science 4 (1977), pp. 277–299

[80] C. Rackoff: *The Covering and Boundedness Problem for Vector Addition Systems.* Theoretical Computer Science 6 (1978), pp. 223–231

[81] E. W. Mayr: *The Complexity of the Finite Containment Problem for Petri Nets.* Cambridge, Mass., MIT Lab. for Computer Science, Technical Report 181 (1977)

[82] E. W. Mayr, A. R. Meyer: *The Complexity of the Finite Containment Problem for Petri Nets.* Journal of the ACM 28, 3 (1981), pp. 561–576

[83] M. Jantzen, H. Bramhoff: *Notions of Computability by Petri Nets,* in [19].

(g) Petri Net Languages

Much effort was spent in the 70ies on the investigation of Net Languages (assign to each transition a character or the empty word and consider sequences of transition firings). Typical papers are e.g. [75], and

[84] M. Hack: *Petri Net Languages.* Computation Structures Group Memo 124, Project MAC, MIT (1975). Also: Technical Report 159, Laboratory for Computer Science MIT Cambridge, Massachusetts (1976)

[85] J. L. Peterson: *Computation Sequence Sets.* Journal of Computer and System Sciences 13, 1 (1976), pp. 1–24

[86] R. Valk, G. Vidal-Naquet: *Petri Nets and Regular Languages.* Journal of Computer and System Sciences 23 (1981), pp. 229–325

[87] S. Crespi-Reghizzi, D. Mandrioli: *Petri Nets and Szilard Languages.* Information and Control 33, No. 2 (1977), pp. 177–192

[88] J. Grabowski: *The Unsolvability of Some Petri Net Language Problems.* Information Processing Letters 9, No. 2 (1979), pp. 60–63

[89] D. Mandrioli: *A Note on Petri Net Languages.* Information and Control 34, No. 2 (1977), pp. 169–171

[90] P. Starke: *Free Petri Net Languages.* Seventh Symposium on Mathematical Foundations of Computer Science 1978, Lecture Notes in Computer Science 64, Springer-Verlag (1978), pp. 506–515

[91] Matthias Jantzen: *On the Hierarchy of Petri Net Languages.* R.A.I.R.O. Informatique théoretique/Theoretical Informatics Vol. 19, No. 1 (1979), pp. 19–30

[92] T. Araki, T. Kagimasa, N. Tokura: *Relations of Flow Languages to Petri Net Languages.* Theoretical Computer Science 15 (1981), pp. 51–75.

This topic is also discussed in the books of Peterson and Starke [21, 23].

(h) Behaviour Representation

As causal dependency and concurrency of transition firings are not represented in firing sequences, several other methods have been suggested to represent the behaviour of place/transition nets:

[93] P. Starke: *Processes in Petri Nets.* Elektronische Informationsverarbeitung und Kybernetik, EIK 17 8/9 (1981), pp. 389–416

[94] G. Rozenberg, R. Verraedt: *Subset Languages for Petri Nets.* Part I: *The Relationship to String Languages and Normal Forms.* Part II: *Closure Properties.* Theoretical Computer Science (1983), Vol. 26, pp. 301–326 and Vol. 27, pp. 85–108

[95] H. D. Burkhard: *Ordered Firing in Petri Nets.* Elektronische Informationsverarbeitung und Kybernetik (EIK) 2/3 (1983), pp. 71–86.

[96] U. Goltz, W. Reisig: *The Non-sequential Behaviour of Petri Nets.* Information & Control, Vol. 57, Nos. 2–3 (1983), pp. 125–147.

The infinite behaviour of place/transition-nets is studied in the paper

[97] R. Valk: *Infinite Behaviour of Petri Nets.* Theoretical Computer Science 25, (3) (1983), pp. 342–373.

9. References to Chapter 6

S-Invariants and *T*-Invariants were introduced by K. Lautenbach in [48]. An overview of more net properties which can be derived by linear algebraic techniques is given in

[98] J. Sifakis: *Structural Properties of Petri Nets.* Mathematical Foundations of Computer Science, Lecture Notes in Computer Science 64, Springer-Verlag (1978), pp. 474–483

and in

[99] G. Memmi, G. Roucairol: *Linear Algebra in Net Theory,* in [17].

Detailed considerations are also contained in the first volume of [22] and in

[100] J. Martinez, M. Silva: *A Simple and Fast Algorithm to obtain all Invariants of a Generalized Petri Net,* in [18].

The seat reservation system of Chap. 6.5 was constructed by Kurt Lautenbach (private communication) and is based on an example by E. Ashcroft.

10. References to Chapter 7

(a) Free Choice Nets

As already mentioned above, M. Hack introduced free choice nets in [7]. Errata to this are collected in

[101] M. Hack: *Corrections to "Analysis of Production Schemata by Petri Nets".* Computation Structures Group Note 17, Project MAC (June 1974).

In [7], Hack proves the deadlock/trap criterion for the liveness of free choice nets. Our proof is a slight modification of his. Further studies on free choice nets include:

[102] E. Best, M. W. Shields: *Some Equivalence Results for Free Choice Nets and Simple Nets and on the Periodicity of Live Free Choice Nets.* Preprint of CAAP 83, 8th Colloquium on Trees in Algebra and Programming, L'Aquila. Lecture Notes in Computer Science 159, Springer-Verlag (1983), pp. 141–154

[103] K. Döpp: *Zum Hackschen Wohlformungssatz für Free-Choice-Petrinetze.* EIK 19, 1/2 (1983), pp. 3–15

Generalizations of the liveness criterion for free choice nets are found in the following two papers:

[104] M. Hack: *Extended State Machine Allocatable Nets (ESMA), an Extension of Free Choice Petri Nets Results,* Computation Structures Group Memo 78, Project MAC, MIT Cambridge, Massachusetts (1973), revised as Memo 78-1 (1974)

[105] W. Griese: *Liveness in NSC-Petri Nets,* in: Discrete Structures and Algorithms, U. Pape (ed.), Carl Hanser Verlag, München (1980)

[106] P. S. Thiagarajan, K. Voss: *A Fresh Look at Free Choice Nets.* Arbeitspapiere der GMD, Nr. 58, October 1983

[107] E. Best, K. Voss: *Free Choice Systems have Home States.* Acta Informatica 21 (1984), pp. 89–100

Similar results on further net classes are discussed in [47]. "Bipolar Schemata" may be considered as a special class of free choice nets:

[108] H. J. Genrich, P. S. Thiagarajan: *A Theory for Bipolar Synchronization Schemes.* Theoretical Computer Science 30 (1984), pp. 241–318

They are also mentioned in [29].

(b) Marked Graphs

The first study on marked graphs was undertaken by H. Genrich in

[109] H. Genrich: *Das Zollstationenproblem.* Internal Reports GMD-I5/69-01-15 and /71-10-13, Gesellschaft für Mathematik und Datenverarbeitung, Bonn (1969 and 1971),

immediately followed by

[110] A. W. Holt, F. Commoner: *Events & Conditions.* Applied Data Research, New York (1970).

Our proofs in Chap. 7.3 are taken from Genrich's paper [109]. More detailed investigations are given in [57] and in the following papers:

[111] F. Commoner, A. W. Holt, S. Even, A. Pnueli: *Marked Directed Graphs.* Journal of Computer and System Sciences 5 (1971), pp. 511–523

[112] H. J. Genrich, K. Lautenbach: *Synchronisationsgraphen.* Acta Informatica 2 (1973), pp. 143–161.

(c) Further Net Classes

Co-ordination of sequential processes is modelled by the following classes of nets:

[113] O. Herzog: *Static Analysis of Concurrent Processes for Dynamic Properties Using Petri Nets.* Lecture Notes in Computer Science 70, Springer-Verlag (1980)

[114] W. Reisig: *Deterministic Buffer Synchronization of Sequential Processes.* Acta Informatica 18 (1982), pp. 117–134

[115] K. Lautenbach, P. S. Thiagarajan: *Analysis of a Resource Allocation Problem Using Petri Nets.* First European Conference on Distributed Processing, Toulouse, J. Syre (ed.), 1979, pp. 260–266

[116] F. De Cindio, G. de Michelis, L. Pomello, C. Simone: *Superposed Automata Nets,* in [18].

There have been investigations trying to find net classes with more or less simple decision procedures for liveness. [8] introduced a class called "simple". They are also studied in [104]. Landweber and Robertson [53] consider "conflict free" nets.

11. References to Chapter 8

An early paper on nets with individual tokens is

[117] M. Schiffers, H. Wedde: *Analyzing Program Solutions of Coordinated Problems by CP-Nets.* Mathematical Foundations of Computer Science 1978, Lecture Notes in Computer Science 64 (1978), pp. 462–473

A fundamental step was the introduction of variables as arc labels in the model of predicate/transition-nets which was introduced in [10]. As a special case of this model one may consider the predicate/event-nets which we introduced in Chap. 8.

A further study of this model is

[118] H. Genrich, K. Lautenbach: *S-Invariance in Predicate/Transition Nets*, in [19].

The distributed database example of Chap. 8.3 is taken from [29].

12. References to Chapter 9

Because the variables in predicate/transition-nets yield difficulties when constructing a calculus of invariants, K. Jensen defined a variable free calculus of nets with individual tokens in

[119] K. Jensen: *Coloured Petri Nets and the Invariant Method.* Theoretical Computer Science 14 (1981), pp. 317–336.

More on this model can be found in

[120] K. Jensen: *How to Find Invariants for Coloured Petri Nets.* Mathematical Foundations of Computer Science 1981, Lecture Notes in Computer Science 118 (1981), pp. 327–338

and in

[121] K. Jensen: *High Level Petri Nets*, in [19].

Relation nets are related to other net models in

[122] W. Reisig: *Petri Nets with Individual Tokens*, in [19].

In FIFO-nets tokens are assumed to be characters, and *S*-elements behave according to the first-in-first-out-principle (hence markings can be considered as character strings):

[123] R. Martin, G. Memmi: *Specification and validation of Sequential Processes Communicating by FIFO Channels.* 4th International Conference of Software Engineering for Telecommunication Switching Systems. (IEEE) Worwick 1981

[124] A. Finkel: *Blocage et vivacité dans les réseaux a pile-file.* STACS 84, Lecture Notes in Computer Science 166 (1984), pp. 151–162.

13. Modifications and Generalizations of Place/Transition-Nets

It is often proposed to modify or to generalize the standard firing rule of place/transition nets or to supply nets with additional components and distinguished interpretations. Most of these generalizations refer to the fact that in place/transition nets it is not possible to test the emptyeness of a place with infinite capacity.

Typical such modifications, as inhibitor arcs and priority rules, are extensively discussed in the books [21, 22, 23].

Evaluation Nets and Macro-*E*-Nets introduce additional types of places:

[125] J. D. Noe, G. J. Nut: *Macro-E-nets for Representations of Parallel Systems.* IEEE Transactions on Computers, Vol. C-22, No. 8 (1973)

[126] J. D. Noe: *Nets in Modelling and Simulation*, in [17].

The concept of dynamic change of arc weights (self modifying nets) is found in

[127] R. Valk: *Generalizations of Petri Nets.* Mathematical Foundations of Computer Science 1981, Lecture Notes in Computer Science 118 (1981), pp. 140–155.

This paper gives also an overview of several net models, their modifications and their properties.

Different types of nets are also compared in

[128] K. Jensen: *A Method to Compare the Descriptive Power of Different Types of Petri Nets.* Mathematical Foundations of Computer Science 1980, Lecture Notes in Computer Science 88, Springer-Verlag (1980), pp. 348–361

[129] S. Porat, M. Yoeli: *Towards a Hierarchy of Nets.* Technion-Israel Institute of Technology, Dept of Computer Science Technical Report No. 224 (1981).

Notions of time are introduced in the following papers:

[130] C. Ramchandani: *Analysis of Asynchronous Concurrent Systems by Petri Nets.* Technical Report 120, Project MAC, MIT Cambridge, Massachusetts (1974)

[133] J. Skifakis: *Performance Evaluation of Systems Using Nets* in [17]
Dept. of Information and Computer Science, University of California, Irvine, California (1974)

[132] S. Ghosh: *Some Comments on Time in Petri Nets* in [16]

[133] J. Skifakis: *Performance Evaluation of Systems Using Nets* in [17]

[134] W. M. Zuberek: *Timed Petri Nets and Preliminary Performance Evaluation.* Proceedings of the 7th Annual Symposium on Computer Architecture, May 6–8, 1980, La Baule, France (1980), pp. 88–96.

Further modifications are given in

[135] M. Moalo, J. Poulou, J. Skifakis: *Synchronized Petri Nets: A Model for the Description of Non-Autonomous Systems.* Mathematical Foundations of Computer Science 1978, Lecture Notes in Computer Science 64, Springer-Verlag (1978), pp. 374–384

[136] M. Yoeli, Z. Barzilai: *Behavioural Descriptions of Communication Switching Systems using Extended Petri Nets.* Digital Processes 3 (1977), pp. 307–320

[137] H. D. Burkhard: *On Priorities of Parallelism: Petri Nets under the Maximum Firing Strategy.* Logics of Programs and their Applications, Lecture Notes in Computer Science 148 (1982)

[138] A. Pistorello, C. Romoli, S. Crespi-Reghizzi: *Threshold Nets and Cell-Assemblies.* Information and Control 49 (1982), pp. 239–264

[139] H. D. Burkhard: *Control of Petri Nets by Finite Automata.* Fundamenta Informaticae Series IV, No. 2, Warszawa (1973)

[140] T. Etzion and M. Yoeli: *Super Nets and Their Hierarchy.* Theoretical Computer Science 25, (2) (1983).

As more general and abstract models one might consider transition systems and subsitution systems:

[141] R. M. Keller: *Vector Replacement Systems: A Formalism for Modelling Asynchronous Systems.* Technical Report 117 Computer Science Laboratory, Princeton University, Princeton, New Jersey (December 1972), revised January 1974

[142] H. J. Genrich, K. Lautenbach, P. S. Thiagarajan: *Substitution Systems – A Family of System Models based on Concurrency.* Mathematical Foundations of Computer Science 1980, Lecture Notes in Computer Science 88, Springer-Verlag (1980), pp. 698–723

[143] J. Sifakis: *A Unified Approach for Studying the Properties of Transition Systems. Theoretical* Computer Science 18 (1982), pp. 227–258.

14. Applications

In this book we presented a few examples of applying nets in system modelling and analysis. Hints on applications in system modelling are also found in the books [20, 22, 23]. A broader spectrum of applications and implementations is contained in the second volume of [22]. Applications are also found in the volumes [17, 18, 19].

Early applications include R. Shapiro's and H. Saint's translation of Fortran programs into nets, showing precedence constraints between operations, and J. Noe's net model of the SCOPE 3.2 operating system:

[144] R. Shapiro, H. Saint: *A New Approach to Optimization of Sequential Decisions.* Annual Review in Automatic Programming, Volume 6, Part 5 (1970), pp. 257–288

[145] J. Noe: *A Petri Net Model for the CDC 6400.* Proceedings ACM SIGOPS Workshop on System Performance Evaluation, New York, ACM (1971), pp. 362–378.

General remarks on the adequate style of net interpretations are made in

[146] C. A. Petri: *Interpretations of Net Theory.* Gesellschaft für Mathematik und Datenverarbeitung, Bonn Internal report ISF-75-07 (1975)

[147] C. A. Petri: *Modelling as a Communication Discipline,* in: H. Beilner, E. Gelembe (eds.): Measuring, Modelling and Evaluation Computer Systems, North-Holland Publ. Comp. (1977), pp. 435–449

[148] C. A. Petri: *Concurrency as a Basis of System Thinking.* Gesellschaft für Mathematik und Datenverarbeitung, Bonn Internal report ISF-78-06 (178) also in Proceedings from 5th Scandinavian Logic Symposium, 1979, Aalborg. F. Jensen, B. Mayoh, K. Moller (eds.), Universitetsforlag Aalborg (1979), pp. 143–162

[149] A. W. Holt: *Net Models of Organizational Systems in Theory and Practice,* in [15].

[150] R. M. Shapiro: *Towards a Design Methodology for Information Systems,* in [15]

[151] C. A. Petri: *Some Personal Views in Net Theory,* in [19].

In the following we survey some of the most prominent application areas of nets:

(a) Hardware

Hardware components are modelled in [125, 136] and in the following papers:

[152] S. Wendt: *Petri-Netze und asynchrone Schaltwerke.* Elektronische Rechenanlagen 16 (1974), pp. 208–216

[153] W. Huen, D. Siewiorek: *Intermodule Protocol for Register Transfer Level Modules: Representation and Analytic Tools.* Proceedings of the Second Annual Symposium on Computer Architecture, New York (1975), pp. 56–62

[154] Kwan Chi Leung, C. Michel, P. Le Beux: *Logical Systems Design Using PLAs and Petri Nets – Programmable Hardwired Systems.* Information Processing 77, B. Gilchrist (ed.), IFIP, North-Holland Publ. Comp. (1977)

[155] J. Grabowski: *On the Analysis of Switching Circuits by Means of Petri Nets.* Elektronische Informationsverarbeitung und Kybernetik (EIK) 14, No. 12 (1978), pp. 611–617

[156] K. Zuse: *Petri Nets from the Engineer's Viewpoint,* in [17]

[157] C. Chaudouard, J. P. Elloy: *A Real Time Monitor and its Representation by Petri Nets.* Micro-processing and Microprogramming 7, North-Holland Publ. Comp. (1981), pp. 241–248

[158] M. Morganti: *Petri-Net Implementation of Recovery Strategies in a large ESS,* in [18]

[159] W. M. Zuberek: *Application of Timed Nets to Analysis of Multiprocessor Realizations of Digital Filters.* Proc. 25th Symposium on Circuits and Systems, Houghton, Michigan, August 1982

[160] W. Kluge, K. Lautenbach: *The Orderly Resolution of Memory Access Conflicts among Competing Channel Processes.* IEEE-Transactions on Computers, vol. 31 (1982), pp. 194–207

(b) Performance Evaluation

Performance evaluation is considered in [125, 133, 134] and in the following papers:

[161] J. Sifakis: *Use of Petri Nets for Performance Evaluation*, in Measuring, Modelling and Evaluating Computer Systems, H. Beilner and E. Gelenbe (eds.), North Holland (1977) pp. 75–93

[162] M. Silva: *Evaluation des Performances des Applications Temps Reel de Type Logique,* in Eighth International Society for Mini- and Micro-Computers, M. H. Hamza (ed.), Acta Press, Anaheim, Calgary, Zurich (1979), pp. 152–157

[163] C. V. Ramamoorthy, G. S. Ho: *Performance Evaluation of Asynchronous Concurrent Systems Using Petri Nets.* IEEE Transactions on Software Engineering Vol. SE-6, No. 5 (1980), pp. 440–449

[164] J. Magott: *Performance Evaluation of Concurrent Systems Using Petri Nets.* Information Processing Letters 18 (1984), pp. 7–13.

(c) Distributed Software Systems

A Distributed Database Scheme is discussed in [10] and in [29]. Further such models are described in

[165] K. Voss: *Using Predicate/Transition-Nets to Model and Analyze Distributed Database Systems.* IEEE Transactions on Software Engineering, Vol. SE-6, No. 6 (1980), pp. 539–544

[166] G. Richter: *IML-Inscribed Nets for Modeling Text Processing and Data(base) Management Systems.* Proceedings of the 7th International Conference on Very Large Data Bases, Cannes (1981), IEEE, pp. 363–375

[167] K. Voss: *Nets as a Consistent Formal Tool for the Stepwise Design and Verification of a Distributed System.* IFIP TC-8 Working Conference on Evolutionary Information Systems, Budapest (1981), J. Hawgood (ed.): Evolutionary Information Systems North Holland (1982), pp. 173–191

[168] P. Rolin: *Using Petri-Nets in Measurement of a Distributed Data Base System,* in [18]

[169] S. Yau, M. U. Caglayan: *Distributed Software System Design Representation Using Modified Petri Nets.* IEEE Transactions on Software Engineering Vol. SE-9, No. 6 (1983), pp. 733–745

(d) Programming Languages

In the following papers nets are used to describe – at least partially – the semantics of programming- and specification-languages.

[170] G. Roucairol: *Une Transformation de Programmes Sequentielles en Programmes Parallèles,* Collèque sur la programmation, Paris 1974. Lecture Notes in Computer Science 19 (1974), pp. 327–349

[171] K. Jensen, M. Kyng, O. L. Madsen: *Delta Semantics Defined by Petri Nets.* University of Aarhus (Denmark) Internal Report PB-95, ISSN 0105-8517 (1979)

[172] P. E. Lauer, P. R. Torrigiani, M. W. Shields: *COSY – A System Specification Language Based on Paths and Processes.* Acta Informatica 12 (1979), pp. 109–158

[173] P. Hruschka, A. Kappatsch, U. Kastens: *Net Attributed Grammars.* University of Karlsruhe (Germany), Institut für Informatik, Internal Report 16/90 (1980)

[174] K. Jensen, M. Kyng: EPSILON, *A System Description Language.* University of Aarhus (Denmark), Internal Report DAIMI PB-150, ISSN 0105-8517 (1982)

[175] N. D. Hansen, K. H. Madsen: *Formal Semantics by a Combination of Denotational Semantics and High Level Petri Nets,* in [19]

[176] M. Kyng: *Specification and Verification of Networks in a Petri Net Based Language,* in [19]

[177] W. E. Kluge, H. Schlüter: *Petri Net Models for the Evaluation of Applicative Programs Based on λ-Expressions.* IEEE-Transactions on Software Engineering, Vol. SE-9, No. 4 (1983), pp. 415–427

(e) Communication Protocols

In recent times Petri Nets have been very successfully applied to modelling and analysis of communication protocols. Some few papers in this area are:

[178] P. Merlin: *A Methodology for Design and Implementation of Communication Protocols.* IEEE Transactions on Computers, Vol. 24, 6 (1976)

[179] C. Girault: *Proof of protocols in case of failures.* Advanced Course on Parallel Processing, University of Loughborough, 1980, D-J. Evens (ed.), Parallel Processing Systems, Press of Cambridge University Press (1980)

[180] J. L. Baer, G. Gardarin, C. Girault, G. Roucairol: *The Two Step Commitment Protocol: Modelling, Specification and Proof Methodology.* 5th international Conference an Software Engineering, San Diego (1981)

[181] M. Diaz: *Modelling and Analysis of Communication and Cooperation Protocols Using Petri Net Based Models.* Tutorial Paper Second International Workshop on Protocol Specification, Testing and Verification May 17–20, 1982. Idyllwild – Los Angeles

[182] F. J. W. Symons: *Representation Analysis and Verification of Communication Protocols.* Telecom Australia Research Laboratories, Victoria, Australia, Report 7380 (1980)

[183] G. Berthelot, R. Terrat: *Petri Net Theory for the Correctness of Protocols.* IEEE Transactions on Computers, C-30 (1982), pp. 2497–2505

[184] P. Estallier, C. Girault: *Petri Nets Specification of a New Protocol for Controlling a Distributed System Organization.* Third International Conference on Distributed Computing Systems Miami, Florida (1982)

[185] P. Estallier, C. Girault: *Petri Net Specification of Virtual Ring Protocols,* in [19].

(f) Further Concepts in Net Applications

There exist applications of nets which are somewhat unexpected, e.g. the net representation of predicate logic in [45]. The interaction among participants which are involved in a lawsuit is represented in

[186] J. Meldman, A. Holt: *Petri Nets and Legal Systems.* Jurimetrics Journal Vol. 12, No. 2 (1971), pp. 65–75.

Other applications of this style are:

[187] H. Genrich: *The Petri Net Representation of Mathematical Knowledge.* Gesellschaft für Mathematik und Informatik, Bonn Internal report SID-76-05 (1976)

[188] A. W. Holt: *Introduction to Occurrence Systems,* in: Associative Information Techniques, L. Jacks (ed.), Elsevier Publishing Company (1971)

[189] H. Oberquelle: *Nets as a Tool in Teaching and in Terminology Work*, in [17]

[190] M. Jantzen: *Structured Representation of Knowledge by Petri Nets as an Aid for Teaching and Research,* in [17]

[191] W. Reisig: *A Note on the Representation of Finite Tree Automata.* Information Processing Letters 8, No. 5 (1979), pp. 239–240.

15. Implementation and Automatic Analysis of Nets

Early Papers on net implementations are

[192] F. Grandoni, P. Zerbetto: *Description and Asynchronous Implementation of Control Structures for Concurrent Systems.* International Computing Symposium 1973, A. Günther et al. (eds.), North-Holland Publ. Comp. (1974), pp. 151–164

and

[193] H. A. Schmid: *An Approach to the Communication and Synchronization of Processes.* International Computing Symposium 1973, A. Günther et al. (eds.), North-Holland Publ. Comp. (1974), pp. 165–171.

Further methods for implementing Petri Nets are discussed in the following papers:

[194] M. Auguin, F. Boeri, C. André: *Systematic Method of Realization of Interpreted Petri Nets.* Digital Processes 6 (1980), pp. 55–68

[195] A. A. Törn: *Simulation Graphs: A General Tool for Modeling Simulation Designs.* Simulation, December 1981, pp. 187–194

[196] G. Berger, G. Florin, S. Natkin: *A Tool for the Dependability and Performance Evaluation of Data Processing Systems.* AFCET Symposium on Mathematics for Computer Science, Paris 1982

[197] J. P. Queille: *The CESAR System: An Aided Design and Certification System for Distributed Applications.* Second International Conference on Distributed Computing Systems, Paris 1981, IEEE, Computer Society Press (1981)

[198] R. A. Nelson, L. M. Haibt, P. B. Sheridan: *Casting Petri Nets into Programs.* IEEE Transactions of Software Engineering, Vol. SE-9, No. 5 (1983), pp. 590–602

Currently a lot of software tools for Petri Net analysis are being developed. A survey of 26 such projects is given in Newsletter 16 of the Special Interest Group on Petri Nets and Related System Models, c.f. [26].

[199] U. Golze, L. Priese: *Petri Net Implementation by a Universal Cell Space.* Information & Control 53 (1982), pp. 121–138.

16. Related System Models

Here we give some pointers to system models which are also used, as are Petri Nets, to describe concurrent systems.

An actual bibliography on this topic is

[200] D. Bell, J. Kerridge, D. Simpson, N. Willis: *Parallel Programming – A bibliography.* Monographs in Informatics Series – Wiley Heyden Ltd.

(a) Papers comparing different models

[201] T. Kasai, R. E. Miller: *Homomorphisms between Models of Parallel Computation.* Journal of Computer and System Sciences 25 (1982), pp. 285–331

[202] R. J. Lipton, L. Syndex, Y. Zalcstein: *A Comparative Study of Models of Parallel Computation.* Proceedings of the 15th Annual Symposium on Switching and Automata Theory, New York, IEEE (1974)

[203] J. Peterson, T. Breth: *A Comparison of Models of Parallel Computation.* Information Processing 74, Proceedings of the 1974, IFIP Congress, Amsterdam (1974), pp. 466−470

[204] J. Baer: *A survey of Some Theoretical Aspects of Multiprocessing.* Computing Surveys 5, Nr. 1 (1973)

[205] R. Miller: *A Comparison of Some Theoretical Models of Parallel Computation.* IEEE Transactions on Computers, Vol. C-22, Nr. 8 (1973), pp. 710−717

[206] R. Miller: *Some Relationships Between Various Models of Parallelism and Synchronization.* Report RC 5074 IBM T. J. Watson Research Center, Yorktown Heights (1974)

[207] F. de Cindio, G. de Michelis, L. Pomello, C. Simone: *Milner's Communicating Systems and Petri Nets,* in [19]

[208] U. Goltz, M. Mycroft: *On the Relationship of CCS and Petri Nets.* ICALP 84 Lecture Notes in Computer Science 172 (1984), pp. 196−208.

(b) Related Models

Quite close to Petri Nets are, of course, the generalizations discussed above in (13). Also, the papers [33−38] describe models which are closely related to nets.

Further Models include:

[209] E. Conry, J. R. Jump: *On Functional Equivalence in a Model for Parallel Computation.* Information & Control 41 (1979), pp. 247−274

[210] R. Karp, R. Miller: *Properties of a Model for Parallel Computation: Determinacy, Termination and Queuing.* SIAM Journal of Applied Mathematics 14, No. 6 (1966), pp. 1390−1411

[211] E. W. Dijkstra: *Cooperating Sequential Processes,* in F. Genuys (editor): Programming Languages, New York, Academic Press (1968)

[212] R. Keller: *Formal Verification of Parallel Programs.* Communications of the ACM, 19, No. 7 (1976), pp. 371−384

[213] G. Kahn, D. MacQueen: *Coroutines and Networks of Parallel Processes,* IFIP 77, Information Processing Conference, B. Gilchrist (editor), North Holland Publ. Company (1977), pp. 993−998

[214] C. A. R. Hoare: *Communicating Sequential Processes.* Communications of the ACM 21, No. 8 (1978), pp. 666−677

[215] R. Milner: *A Calculus of Communicating Systems.* Lecture Notes in Computer Science 92 (1980)

[216] A. Maggiolo-Schettini, H. Wedde, J. Winkowski: *Modelling a Solution for a Control in Distributed Systems by Restrictions.* Theoretical Computer Science 13 (1981), pp. 61−83

[217] J. W. de Bakker, J. I. Zucker: *Processes and the Denotational Semantics of Concurrency.* Information and Control 54 (1/2 July/August 1982), pp. 70−120

[218] L. Priese: *Automata and Concurrency.* Theoretical Computer Science 25 (1983), pp. 221−265

[219] R. Milner: *Calculi for Synchrony and Asynchrony.* Theoretical Computer Science 25 (1983), pp. 267−310

[220] J. I. Castellani, P. Franceschi, U. Montanari: *Labeled Event Structures: A Model for Observable Concurrency,* in: Formal Description of Programming Concepts II, D. Bjørner (ed.), North-Holland Publ. Comp. IFIP (1983), pp. 383−399

[221] H. J. Genrich, P. S. Thiagarajan: *Well Formed Flow Charts for Concurrent Programming,* in: Formal Description of Programming Concepts-II, D. Bjørner (ed.), North-Holland Publ. Comp. IFIP (1983), pp. 357−380.

Index

Page numbers in *italics* refer to definitions

EATCS Monographs on Theoretical Computer Science

Editors: **W. Brauer, G. Rozenberg, A. Salomaa**

K. Mehlhorn

Data Structures and Algorithms 2 Graph Algorithms and NP-Completeness

1984. 54 figures. XII, 260 pages.
ISBN 3-540-13641-X

Contents: Algorithms on Graphs. – Path Problems in Graphs and Matrix Multiplication. – NP-Completeness. – Algorithmic Paradigms. – Bibliography. – Subject Index.

K. Mehlhorn

Data Structures and Algorithms 3 Multi-dimensional Searching and Computational Geometry

1984. 134 figures. XII, 284 pages.
ISBN 3-540-13642-8

Contents: Multidimensional Data Structures. – Computational Geometry. – Algorithmic Paradigms. – Bibliography. – Subject Index.

Springer-Verlag
Berlin
Heidelberg
New York
Tokyo